THE ALGEBRA OF LOGIC

BY

LOUIS COUTURAT

AUTHORIZED ENGLISH TRANSLATION

BY

LYDIA GILLINGHAM ROBINSON, B. A.

Illustration de couverture : © domaine public
Edition : Culturea, le patrimoine des lettres (Hérault, 34)
Contact : infos@culturea.fr
Retrouvez notre catalogue sur http://culturea.fr
Imprimé en Allemagne par Books on Demand, In de Tarpen 42, Norderstedt.
Design typographique : Derek Murphy
Layout : Reedsy (https://reedsy.com/)
ISBN : 9791041940806
Dépôt légal : janvier 2023

Preface

Mathematical Logic is a necessary preliminary to logical Mathematics. "Mathematical Logic" is the name given by PEANO to what is also known (after VENN) as "Symbolic Logic"; and Symbolic Logic is, in essentials, the Logic of Aristotle, given new life and power by being dressed up in the wonderful—almost magical—armour and accoutrements of Algebra. In less than seventy years, logic, to use an expression of DE MORGAN'S, has so *thriven* upon symbols and, in consequence, so grown and altered that the ancient logicians would not recognize it, and many old-fashioned logicians will not recognize it. The metaphor is not quite correct: Logic has neither grown nor altered, but we now see more *of* it and more *into* it.

The primary significance of a symbolic calculus seems to lie in the economy of mental effort which it brings about, and to this is due the characteristic power and rapid development of mathematical knowledge. Attempts to treat the operations of formal logic in an analogous way had been made not infrequently by some of the more philosophical mathematicians, such as LEIBNIZ and LAMBERT ; but their labors remained little known, and it was BOOLE and DE MORGAN, about the middle of the nineteenth century, to whom a mathematical—though of course non-quantitative—way of regarding logic was due. By this, not only was the traditional or Aristotelian doctrine of logic reformed and completed, but out of it has developed, in course of time, an instrument which deals in a sure manner with the task of investigating the fundamental concepts of mathematics—a task which philosophers have repeatedly taken in hand, and in which they have as repeatedly failed.

First of all, it is necessary to glance at the growth of symbolism in mathematics; where alone it first reached perfection. There have been three stages in the development of mathematical doctrines: first came propositions with particular numbers, like the one expressed, with signs subsequently invented, by "$2+3=5$"; then came more general laws holding for all numbers and expressed by letters, such as

$$\text{“}(a+b)c = ac+bc\text{”;}$$

lastly came the knowledge of more general laws of functions and the formation of the conception and expression "function". The origin of the symbols for particular whole numbers is very ancient, while the symbols now in use for the operations and relations of arithmetic mostly date from the sixteenth and seventeenth centuries; and these "constant" symbols together with the letters first used systematically by VIÈTE (1540–1603) and DESCARTES (1596–1650), serve, by themselves, to express many propositions. It is not, then, surprising that DESCARTES, who was both a mathematician and a philosopher, should have had the idea of keeping the method of algebra while going beyond the material of traditional mathematics and embracing the general science of what thought finds, so that philosophy should become a kind of Universal Mathematics. This sort of generalization of the use of symbols for analogous theories is a characteristic of mathematics, and seems to be a reason lying deeper than the erroneous idea, arising from a simple confusion of thought, that algebraical symbols necessarily imply something quantitative, for the antagonism there used to be and is on the part of those logicians who were not and are not mathematicians, to symbolic logic. This idea of a universal mathematics was cultivated especially by GOTTFRIED WILHELM LEIBNIZ (1646–1716).

Though modern logic is really due to BOOLE and DE MORGAN, LEIBNIZ was the first to have a really distinct plan of a system of mathematical logic. That this is so appears from research—much of which is quite recent—into LEIBNIZ'S unpublished work.

The principles of the logic of LEIBNIZ, and consequently of his whole philosophy, reduce to two[1]: (1) All our ideas are compounded of a very small number of simple ideas which form the "alphabet of human thoughts"; (2) Complex ideas proceed from these simple ideas by a uniform and symmetrical combination which is analogous to arithmetical multiplication. With regard to the first principle, the number of simple ideas is much greater than LEIBNIZ thought; and, with regard to the second principle, logic considers three operations—which we shall meet with in the following book under the names of logical multiplication, logical addition and negation—instead of only one.

"Characters" were, with LEIBNIZ, any written signs, and "real" characters were those which—as in the Chinese ideography—represent ideas directly, and not the words for them. Among real characters, some simply serve to represent ideas, and some serve for reasoning. Egyptian and Chinese hieroglyphics and the symbols of astronomers and chemists belong to the first category, but LEIBNIZ declared them to be imperfect, and desired the second category of characters for what he called his "universal characteristic".[2] It was not in the form of an algebra that LEIBNIZ first conceived his characteristic, probably because he was then a novice in mathematics, but in the form of a universal language or script.[3] It was in 1676 that he first dreamed of a kind of algebra of thought,[4] and it was the algebraic notation which then served as model for the characteristic.[5]

LEIBNIZ attached so much importance to the invention of proper symbols that he attributed to this alone the whole of his discoveries in mathematics.[6] And, in fact, his infinitesimal calculus affords a most brilliant example of the importance of, and LEIBNIZ' s skill in devising, a suitable notation.[7]

Now, it must be remembered that what is usually understood by the name "symbolic logic", and which—though not its name—is chiefly due to BOOLE, is what LEIBNIZ called a *Calculus ratiocinator*, and is only a part of the Universal Characteristic. In symbolic logic LEIBNIZ enunciated the principal properties of what we now call logical multiplication, addition, negation, identity, class-inclusion, and the null-class; but the aim of LEIBNIZ'S researches was, as he

[1] COUTURAT, *La Logique de Leibniz d'après des documents inédits*, Paris, 1901, pp. 431–432, 48.
[2] *Ibid.*, p. 81.
[3] *Ibid.*, pp. 51, 78
[4] *Ibid.*, p. 61.
[5] *Ibid.*, p. 83.
[6] *Ibid.*, p. 84.
[7] *Ibid.*, p. 84–87.

said, to create "a kind of general system of notation in which all the truths of reason should be reduced to a calculus. This could be, at the same time, a kind of universal written language, very different from all those which have been projected hitherto; for the characters and even the words would direct the reason, and the errors—excepting those of fact—would only be errors of calculation. It would be very difficult to invent this language or characteristic, but very easy to learn it without any dictionaries". He fixed the time necessary to form it: "I think that some chosen men could finish the matter within five years"; and finally remarked: "And so I repeat, what I have often said, that a man who is neither a prophet nor a prince can never undertake any thing more conducive to the good of the human race and the glory of God".

In his last letters he remarked: "If I had been less busy, or if I were younger or helped by well-intentioned young people, I would have hoped to have evolved a characteristic of this kind"; and: "I have spoken of my general characteristic to the Marquis de l'Hôpital and others; but they paid no more attention than if I had been telling them a dream. It would be necessary to support it by some obvious use; but, for this purpose, it would be necessary to construct a part at least of my characteristic;—and this is not easy, above all to one situated as I am".

LEIBNIZ thus formed projects of both what he called a *characteristica universalis*, and what he called a *calculus ratiocinator*; it is not hard to see that these projects are interconnected, since a perfect universal characteristic would comprise, it seems, a logical calculus. LEIBNIZ did not publish the incomplete results which he had obtained, and consequently his ideas had no continuators, with the exception of LAMBERT and some others, up to the time when BOOLE, DE MORGAN, SCHRÖDER, MACCOLL, and others rediscovered his theorems. But when the investigations of the principles of mathematics became the chief task of logical symbolism, the aspect of symbolic logic as a calculus ceased to be of such importance, as we see in the work of FREGE and RUSSELL. FREGE'S symbolism, though far better for logical analysis than BOOLE'S or the more modern PEANO'S, for instance, is far inferior to PEANO'S —a symbolism in which the merits of internationality and power of expressing mathematical theorems are very satisfactorily attained—in practical convenience. RUSSELL, especially in his later works, has used the ideas of FREGE, many of which he discovered subsequently to, but independently of, FREGE, and modified the symbolism of PEANO as little as possible. Still, the complications thus introduced take away that simple character which seems necessary to a calculus, and which BOOLE and others reached by passing over certain distinctions which a subtler logic has shown us must ultimately be made.

Let us dwell a little longer on the distinction pointed out by LEIBNIZ between a *calculus ratiocinator* and a *characteristica universalis* or *lingua characteristica*. The ambiguities of ordinary language are too well known for it to be necessary for us to give instances. The objects of a complete logical symbolism are: firstly, to avoid this disadvantage by providing an *ideography*, in which the signs represent ideas and the relations between them *directly* (without the intermediary of words), and secondly, so to manage that, from given premises, we can, in this ideography, draw all the logical conclusions which they imply by means of rules of transformation of formulas analogous to those of algebra,—in fact, in which we can replace reasoning by the almost mechanical process of calculation. This second requirement is the requirement of a *calculus ratiocinator*. It is essential that the ideography should be complete, that only symbols with a well-defined meaning should be used—to avoid the same sort of ambiguities that words have—and, consequently,—that no suppositions should be introduced implicitly, as is commonly the case if the meaning of signs is not well defined. Whatever premises are necessary and sufficient for a conclusion should be stated explicitly.

Besides this, it is of practical importance,—though it is theoretically irrelevant,—that the ideography should be concise, so that it is a sort of stenography.

The merits of such an ideography are obvious: rigor of reasoning is ensured by the calculus character; we are sure of not introducing unintentionally any premise; and we can see exactly on what propositions any demonstration depends.

We can shortly, but very fairly accurately, characterize the dual development of the theory of symbolic logic during the last sixty years as follows: The *calculus ratiocinator* aspect of symbolic logic was developed by BOOLE, DE MORGAN, JEVONS, VENN, C. S. PEIRCE, SCHRÖDER, Mrs. LADD-FRANKLIN and others; the *lingua characteristica* aspect was developed by FREGE, PEANO and RUSSELL. Of course there is no hard and fast boundary-line between the domains of these two parties. Thus PEIRCE and SCHRÖDER early began to work at the foundations of arithmetic with the help of the calculus of relations; and thus they did not consider the logical calculus merely as an interesting branch of algebra. Then PEANO paid particular attention to the calculative aspect of his symbolism. FREGE has remarked that his own symbolism is meant to be a *calculus ratiocinator* as well as a *lingua characteristica*, but the using of FREGE'S symbolism as a calculus would be rather like using a three-legged stand-camera for what is called "snap-shot" photography, and one of the outwardly most noticeable things about RUSSELL'S work is his combination of the symbolisms of FREGE and PEANO in such a way as to preserve nearly all of the merits of each.

The present work is concerned with the *calculus ratiocinator* aspect, and shows, in an admirably succinct form, the beauty, symmetry and simplicity of the calculus of logic regarded as an algebra. In fact, it can hardly be doubted that some such form as the one in which SCHRÖDER left it is by far the best for exhibiting it from this point of view.[8] The content of the present volume corresponds to the two first volumes of SCHRÖDER'S great but rather prolix treatise.[9] Principally owing to the influence of C. S. PEIRCE, SCHRÖDER

[8] Cf. A. N. WHITEHEAD, *A Treatise on Universal Algebra with Applications*, Cambridge, 1898.

[9] *Vorlesungen über die Algebra der Logik*, Vol. I., Leipsic, 1890; Vol. II, 1891 and 1905. We may mention that a much shorter *Abriss* of the work has been prepared by EUGEN MÜLLER. Vol. III (1895) of SCHRÖDER'S work is on the logic of relatives founded by DE MORGAN and C. S. PEIRCE, —a branch of Logic that is only mentioned in the concluding sentences

departed from the custom of BOOLE, JEVONS, and himself (1877), which consisted in the making fundamental of the notion of *equality*, and adopted the notion of *subordination* or *inclusion* as a primitive notion. A more orthodox BOOLIAN exposition is that of VENN, [10] which also contains many valuable historical notes.

We will finally make two remarks.

When BOOLE (cf. §0.2 below) spoke of propositions determining a class of moments at which they are true, he really (as did MACCOLL) used the word "proposition" for what we now call a "propositional function". A "proposition" is a thing expressed by such a phrase as "twice two are four" or "twice two are five", and is always true or always false. But we might seem to be stating a proposition when we say: "Mr. WILLIAM JENNINGS BRYAN is Candidate for the Presidency of the United States", a statement which is sometimes true and sometimes false. But such a statement is like a mathematical *function* in so far as it depends on a *variable*—the time. Functions of this kind are conveniently distinguished from such entities as that expressed by the phrase "twice two are four" by calling the latter entities "propositions" and the former entities "propositional functions": when the variable in a propositional function is fixed, the function becomes a proposition. There is, of course, no sort of necessity why these special names should be used; the use of them is merely a question of convenience and convention.

In the second place, it must be carefully observed that, in §0.13, 0 and 1 are not *defined* by expressions whose principal copulas are relations of inclusion. A definition is simply the convention that, for the sake of brevity or some other convenience, a certain new sign is to be used instead of a group of signs whose meaning is already known. Thus, it is the sign of *equality* that forms the principal copula. The theory of definition has been most minutely studied, in modern times by FREGE and PEANO.

Philip E. B. Jourdain.

Girton, Cambridge. England.

of this volume.

[10] Symbolic Logic, London, 1881; 2nd ed., 1894.

Contents

Bibliography[11]

GEORGE BOOLE. *The Mathematical Analysis of Logic* (Cambridge and London, 1847).

— *An Investigation of the Laws of Thought* (London and Cambridge, 1854).

W. STANLEY JEVONS. *Pure Logic* (London, 1864).

— "On the Mechanical Performance of Logical Inference" (*Philosophical Transactions*, 1870).

ERNST SCHRÖDER. *Der Operationskreis des Logikkalkuls* (Leipsic, 1877).

— *Vorlesungen über die Algebra der Logik*, Vol. I (1890), Vol. II (1891), Vol. III: *Algebra und Logik der Relative* (1895) (Leipsic).[12]

ALEXANDER MACFARLANE. *Principles of the Algebra of Logic, with Examples* (Edinburgh, 1879).

JOHN VENN. *Symbolic Logic*, 1881; 2nd. ed., 1894 (London).[13] *Studies in Logic* by members of the Johns Hopkins University (Boston, 1883): particularly Mrs. LADD-FRANKLIN, O. MITCHELL and C. S. PEIRCE.

A. N. WHITEHEAD. *A Treatise on Universal Algebra*. Vol. I (Cambridge, 1898).

— "Memoir on the Algebra of Symbolic Logic" (*American Journal of Mathematics*, Vol. XXIII, 1901).

EUGEN MÜLLER. *Über die Algebra der Logik:* I. *Die Grundlagen des Gebietekalkuls;* II. *Das Eliminationsproblem und die Syllogistik;* Programs of the Grand Ducal Gymnasium of Tauberbischofsheim (Baden), 1900, 1901 (Leipsic).

W. E. JOHNSON. "Sur la théorie des égalités logiques" (*Bibliothèque du Congrès international de Philosophie*. Vol. III, *Logique et Histoire des Sciences;* Paris, 1901).

PLATON PORETSKY. *Sept Lois fondamentales de la théorie des égalités logiques* (Kazan, 1899).

— *Quelques lois ultérieures de la théorie des égalités logiques* (Kazan, 1902).

— "Exposé élémentaire de la théorie des égalités logiques à deux termes" (*Revue de Métaphysique et de Morale*. Vol. VIII, 1900.)

[11] This list contains only the works relating to the system of BOOLE and SCHRÖDER explained in this work.

[12] EUGEN MÜLLER has prepared a part, and is preparing more, of the publication of supplements to Vols. II and III, from the papers left by SCHRÖDER.

[13] A valuable work from the points of view of history and bibliography.

— "Théorie des égalités logiques à trois termes" (*Bibliothèque du Congrès international de Philosophie*). Vol. III. (*Logique et Histoire des Sciences*). (Paris, 1901, pp. 201–233).

— *Théorie des non-égalités logiques* (Kazan, 1904).

E. V. HUNTINGTON. "Sets of Independent Postulates for the Algebra of Logic" (*Transactions of the American Mathematical Society*, Vol. V, 1904).

THE ALGEBRA OF LOGIC.

0.1 Introduction

The algebra of logic was founded by GEORGE BOOLE (1815–1864); it was developed and perfected by ERNST SCHRÖDER (1841–1902). The fundamental laws of this calculus were devised to express the principles of reasoning, the "laws of thought". But this calculus may be considered from the purely formal point of view, which is that of mathematics, as an algebra based upon certain principles arbitrarily laid down. It belongs to the realm of philosophy to decide whether, and in what measure, this calculus corresponds to the actual operations of the mind, and is adapted to translate or even to replace argument; we cannot discuss this point here. The formal value of this calculus and its interest for the mathematician are absolutely independent of the interpretation given it and of the application which can be made of it to logical problems. In short, we shall discuss it not as logic but as algebra.

0.2 The Two Interpretations of the Logical Calculus

There is one circumstance of particular interest, namely, that the algebra in question, like logic, is susceptible of two distinct interpretations, the parallelism between them being almost perfect, according as the letters represent concepts or propositions. Doubtless we can, with BOOLE and SCHRÖDER, reduce the two interpretations to one, by considering the concepts on the one hand and the propositions on the other as corresponding to *assemblages* or *classes*; since a concept determines the class of objects to which it is applied (and which in logic is called its *extension*), and a proposition determines the class of the instances or moments of time in which it is true (and which by analogy can also be called its extension). Accordingly the calculus of concepts and the calculus of propositions become reduced to but one, the calculus of classes, or, as LEIBNIZ called it, the theory of the whole and part, of that which contains and that which is contained. But as a matter of fact, the calculus of concepts and the calculus of propositions present certain differences, as we shall see, which prevent their complete identification from the formal point of view and consequently their reduction to a single "calculus of classes".

Accordingly we have in reality three distinct calculi, or, in the part common to all, three different interpretations of the same calculus. In any case the reader must not forget that the logical value and the deductive sequence of the formulas does not in the least depend upon the interpretations which may be given them, and, in order to make this necessary abstraction easier, we shall take care to place the symbols "C. I." (*conceptual interpretation*) and "P. I." (*prepositional interpretation*) before all interpretative phrases. These interpretations shall serve only to render the formulas intelligible, to give them clearness and to make their meaning at once obvious, but never to justify them. They may be omitted without destroying the logical rigidity of the system.

In order not to favor either interpretation we shall say that the letters represent *terms*; these terms may be either concepts or propositions according to the case in hand. Hence we use the word *term* only in the logical sense. When we wish to designate the "terms" of a sum we shall use the word *summand* in order that the logical and mathematical meanings of the word may not be confused. A term may therefore be either a factor or a summand.

0.3 Relation of Inclusion

Like all deductive theories, the algebra of logic may be established on various systems of principles[14]; we shall choose the one which most nearly approaches the exposition of SCHRÖDER and current logical interpretation.

The fundamental relation of this calculus is the binary (two-termed) relation which is called *inclusion* (for classes), *subsumption* (for concepts), or *implication* (for propositions). We will adopt the first name as affecting alike the two logical interpretations, and we will represent this relation by the sign $<$ because it has formal properties analogous to those of the mathematical relation $<$ ("less than") or more exactly $\leq$, especially the relation of not being symmetrical. Because of this analogy SCHRÖDER represents this relation by the sign $\in$ which we shall not employ because it is complex, whereas the relation of inclusion is a simple one.

In the system of principles which we shall adopt, this relation is taken as a primitive idea and is consequently indefinable. The explanations which follow are not given for the purpose of *defining* it but only to indicate its meaning according to each of the two interpretations.

C. I.: When a and b denote concepts, the relation $a < b$ signifies that the concept a is subsumed under the concept b; that is, it is a species with respect to the genus b. From the extensive point of view, it denotes that the class of a's is contained in the class of b's or makes a part of it; or, more concisely, that "All a's are b's". From the comprehensive point of view it means that the concept b is contained in the concept a or makes a part of it, so that consequently the character a implies or involves the character b. Example: "All men are mortal"; "Man implies mortal"; "Who says man says mortal"; or, simply, "Man, therefore mortal".

P. I.: When a and b denote propositions, the relation $a < b$ signifies that the proposition a implies or involves the proposition b, which is often expressed by the hypothetical judgement, "If a is true, b is true"; or by "a implies b"; or more simply by "a, therefore b". We see that in both interpretations the relation $<$ may be translated approximately by "therefore".

[14]See HUNTINGTON, "Sets of Independent Postulates for the Algebra of Logic", *Transactions of the Am. Math. Soc.*, Vol. V, 1904, pp. 288–309. [Here he says: "Any set of consistent postulates would give rise to a corresponding algebra, viz., the totality of propositions which follow from these postulates by logical deductions. Every set of postulates should be free from redundances, in other words, the postulates of each set should be *independent*, no one of them deducible from the rest."]

Remark.—Such a relation as "$a < b$" is a proposition, whatever may be the interpretation of the terms a and b. Consequently, whenever a $<$ relation has two like relations (or even only one) for its members, it can receive only the propositional interpretation, that is to say, it can only denote an implication.

A relation whose members are simple terms (letters) is called a *primary* proposition; a relation whose members are primary propositions is called a *secondary* proposition, and so on.

From this it may be seen at once that the propositional interpretation is more homogeneous than the conceptual, since it alone makes it possible to give the same meaning to the copula $<$ in both primary and secondary propositions.

0.4 Definition of Equality

There is a second copula that may be defined by means of the first; this is the copular $=$ ("equal to"). By definition we have

$$a = b,$$

whenever

$$a < b \text{ and } b < a$$

are true at the same time, and then only. In other words, the single relation $a = b$ is equivalent to the two simultaneous relations $a < b$ and $b < a$.

In both interpretations the meaning of the copula $=$ is determined by its formal definition:

C. I.: $a = b$ means, "All a's are b's and all b's are a's"; in other words, that the classes a and b coincide, that they are identical.[15]

P. I.: $a = b$ means that a implies b and b implies a; in other words, that the propositions a and b are equivalent, that is to say, either true or false at the same time.[16]

Remark.—The relation of equality is symmetrical by very reason of its definition: $a = b$ is equivalent to $b = a$. But the relation of inclusion is not symmetrical: $a < b$ is not equivalent to $b < a$, nor does it imply it. We might agree to consider the expression $a > b$ equivalent to $b < a$, but we prefer for the sake of clearness to preserve always the same sense for the copula $<$. However, we might translate verbally the same inclusion $a < b$ sometimes by "a is contained in b", and sometimes by "b contains a".

In order not to favor either interpretation, we will call the first member of this relation the *antecedent* and the second the *consequent*.

C. I.: The antecedent is the *subject* and the consequent is the *predicate* of a universal affirmative proposition.

[15]This does not mean that the concepts a and b have the same meaning. Examples: "triangle" and "trilateral", "equiangular triangle" and "equilateral triangle".

[16]This does not mean that they have the same meaning. Example: "The triangle ABC has two equal sides", and "The triangle ABC has two equal angles".

P. I.: The antecedent is the *premise* or the *cause*, and the consequent is the *consequence*. When an implication is translated by a *hypothetical* (or *conditional*) judgment the antecedent is called the *hypothesis* (or the *condition*) and the consequent is called the *thesis*.

When we shall have to demonstrate an equality we shall usually analyze it into two converse inclusions and demonstrate them separately. This analysis is sometimes made also when the equality is a datum (a *premise*).

When both members of the equality are propositions, it can be separated into two implications, of which one is called a *theorem* and the other its *reciprocal*. Thus whenever a theorem and its reciprocal are true we have an equality. A simple theorem gives rise to an implication whose antecedent is the *hypothesis* and whose consequent is the *thesis* of the theorem.

It is often said that the hypothesis is the *sufficient condition* of the thesis, and the thesis the *necessary condition* of the hypothesis; that is to say, it is sufficient that the hypothesis be true for the thesis to be true; while it is necessary that the thesis be true for the hypothesis to be true also. When a theorem and its reciprocal are true we say that its hypothesis is the necessary and sufficient condition of the thesis; that is to say, that it is at the same time both cause and consequence.

0.5 Principle of Identity

The first principle or axiom of the algebra of logic is the *principle of identity*, which is formulated thus:

Ax. 1

$$a < a,$$

whatever the term a may be.

C. I.: "All a's are a's", *i.e.*, any class whatsoever is contained in itself.

P. I.: "a implies a", *i.e.*, any proposition whatsoever implies itself.

This is the primitive formula of the principle of identity. By means of the definition of equality, we may deduce from it another formula which is often wrongly taken as the expression of this principle:

$$a = a,$$

whatever a may be; for when we have

$$a < a, a < a,$$

we have as a direct result,

$$a = a.$$

C. I.: The class a is identical with itself.

P. I.: The proposition a is equivalent to itself.

0.6 Principle of the Syllogism

Another principle of the algebra of logic is the principle of the *syllogism*, which may be formulated as follows:

Ax. 2

$$(a < b)(b < c) < (a < c).$$

C. I.: "If all a's are b's, and if all b's are c's, then all a's are c's". This is the principle of the *categorical syllogism*.

P. I.: "If a implies b, and if b implies c, a implies c." This is the principle of the *hypothetical syllogism*.

We see that in this formula the principal copula has always the sense of implication because the proposition is a secondary one.

By the definition of equality the consequences of the principle of the syllogism may be stated in the following formulas[17]:

$$\begin{aligned} &(a < b) \quad (b = c) < (a < c), \\ &(a = b) \quad (b < c) < (a < c), \\ &(a = b) \quad (b - c) < (a = c). \end{aligned}$$

The conclusion is an equality only when both premises are equalities.

The preceding formulas can be generalized as follows:

$$\begin{aligned} &(a < b) \quad (b < c) \quad (c < d) < (a < d), \\ &(a = b) \quad (b = c) \quad (c = d) < (a = d). \end{aligned}$$

Here we have the two chief formulas of the *sorites*. Many other combinations may be easily imagined, but we can have an equality for a conclusion only when all the premises are equalities. This statement is of great practical value. In a succession of deductions we must pay close attention to see if the transition from one proposition to the other takes place by means of an equivalence or only of an implication. There is no equivalence between two extreme propositions unless all intermediate deductions are equivalences; in other words, if there is one single implication in the chain, the relation of the two extreme propositions is only that of implication.

0.7 Multiplication and Addition

The algebra of logic admits of three operations, *logical multiplication*, *logical addition*, and *negation*. The two former are binary operations, that is to say, combinations of two terms having as a consequent a third term which may or may not be different from each of them. The existence of the logical product

[17]Strictly speaking, these formulas presuppose the laws of multiplication which will be established further on; but it is fitting to cite them here in order to compare them with the principle of the syllogism from which they are derived.

and logical sum of two terms must necessarily answer the purpose of a double postulate, for simply to define an entity is not enough for it to exist. The two postulates may be formulated thus:

Ax. 3 Given any two terms, a and b, then there is a term p such that

$$p < a, p < b,$$

and that for every value of x for which

$$x < a, x < b,$$

we have also

$$x < p.$$

Ax. 4 Given any two terms, a and b, there exists a term s such that

$$a < s, b < s,$$

we have also

$$s < x.$$

It is easily proved that the terms p and s determined by the given conditions are unique, and accordingly we can define *the* product ab and *the* sum $a + b$ as being respectively the terms p and s.

C. I.: 1. The product of two classes is a class p which is contained in each of them and which contains every (other) class contained in each of them;

2. The sum of two classes a and b is a class s which contains each of them and which is contained in every (other) class which contains each of them.

Taking the words "less than" and "greater than" in a metaphorical sense which the analogy of the relation $<$ with the mathematical relation of inequality suggests, it may be said that the product of two classes is the greatest class contained in both, and the sum of two classes is the smallest class which contains both.[18] Consequently the product of two classes is the part that is common to each (the class of their common elements) and the sum of two classes is the class of all the elements which belong to at least one of them.

P. I.: 1. The product of two propositions is a proposition which implies each of them and which is implied by every proposition which implies both:

2. The sum of two propositions is the proposition which is implied by each of them and which implies every proposition implied by both.

Therefore we can say that the product of two propositions is their weakest common cause, and that their sum is their strongest common consequence, strong and weak being used in a sense that every proposition which implies

[18] According to another analogy DEDEKIND designated the logical sum and product by the same signs as the least common multiple and greatest common divisor (*Was sind und was sollen die Zahlen?* Nos. 8 and 17, 1887. [Cf. English translation entitled *Essays on Number* (Chicago, Open Court Publishing Co. 1901, pp. 46 and 48)] GEORG CANTOR originally gave them the same designation (*Mathematische Annalen*, Vol. XVII, 1880).

another is stronger than the latter and the latter is weaker than the one which implies it. Thus it is easily seen that the product of two propositions consists in their *simultaneous affirmation*: "a and b are true", or simply "a and b"; and that their sum consists in their *alternative affirmation*, "either a or b is true", or simply "a or b".

Remark.—Logical addition thus defined is not disjunctive;[19] that is to say, it does not presuppose that the two summands have no element in common.

0.8 Principles of Simplification and Composition

The two preceding definitions, or rather the postulates which precede and justify them, yield directly the following formulas:

(1) $$ab < a, \qquad ab < b,$$

(2) $$(x < a)(x < b) < (x < ab),$$

(3) $$a < a + b, \qquad b < a + b,$$

(4) $$(a < x)(b < x) < (a + b < x).$$

Formulas (1) and (3) bear the name of the *principle of simplification* because by means of them the premises of an argument may be simplified by deducing therefrom weaker propositions, either by deducing one of the factors from a product, or by deducing from a proposition a sum (alternative) of which it is a summand.

Formulas (2) and (4) are called the *principle of composition*, because by means of them two inclusions of the same antecedent or the same consequent may be combined (*composed*). In the first case we have the product of the consequents, in the second, the sum of the antecedents.

The formulas of the principle of composition can be transformed into equalities by means of the principles of the syllogism and of simplification. Thus we have

1 (Syll.) $$(x < ab)(ab < a) < (x < a),$$

(Syll.) $$(x < ab)(ab < b) < (x < b).$$

Therefore

(Comp.) $$(x < ab) < (x < a)(x < b).$$

2 (Syll.) $$(a < a + b)(a + b < x) < (a < x),$$

(Syll.) $$(b < a + b)(a + b < x) < (b < x).$$

[19] BOOLE, closely following analogy with ordinary mathematics, premised, as a necessary condition to the definition of "$x + y$", that x and y were mutually exclusive. JEVONS, and practically all mathematical logicians after him, advocated, on various grounds, the definition of "logical addition" in a form which does not necessitate mutual exclusiveness.

Therefore

(Comp.) $$(a + b < x) < (a < x)(b < x).$$

If we compare the new formulas with those preceding, which are their converse propositions, we may write

$$(x < ab) = (x < a)(x < b),$$
$$(a + b < x) = (a < x)(b < x).$$

Thus, to say that x's contained in ab is equivalent to saying that it is contained at the same time in both a and b; and to say that x contains $a + b$ is equivalent to saying that it contains at the same time both a and b.

0.9 The Laws of Tautology and of Absorption

Since the definitions of the logical sum and product do not imply any order among the terms added or multiplied, logical addition and multiplication evidently possess commutative and associative properties which may be expressed in the formulas

$$ab = ba, \qquad a + b = b + a,$$
$$(ab)c = a(bc), \qquad (a + b) + c = a + (b + c).$$

Moreover they possess a special property which is expressed in the *law of tautology:*

$$a = aa, \qquad a = a + a.$$

Demonstration:

1 (Simpl.) $$aa < a,$$
(Comp.) $$(a < a)(a < a) = (a < aa)$$

whence, by the definition of equality,

$$(aa < a)(a < aa) = (a - aa)$$

In the same way:

2 (Simpl.) $$a < a + a,$$
(Comp.) $$(a < a)(a < a) = (a + a < a),$$

whence

$$(a < a + a)(a + a < a) = (a = a + a).$$

From this law it follows that the sum or product of any number whatever of equal (identical) terms is equal to one single term. Therefore in the algebra of logic there are neither multiples nor powers, in which respect it is very much simpler than numerical algebra.

Finally, logical addition and multiplication posses a remarkable property which also serves greatly to simplify calculations, and which is expressed by the *law of absorption:*

$$a + ab = a, \qquad a(a+b) = a.$$

Demonstration:

1 (Comp.) $\quad (a < a)(ab < a) < (a + ab < a),$

(Simpl.) $\quad a < a + ab,$

whence, by the definition of equality,

$$(a + ab < a)(a < a + ab) = (a + ab = a).$$

In the same way:

1 (Comp.) $\quad (a < a)(a < a + b) < [a < a(a+b)],$

(Simpl.) $\quad a(a+b) < a,$

whence

$$[a < a(a+b)][a(a+b) < a] = [a(a+b) = a].$$

Thus a term (a) *absorbs* a summand (ab) of which it is a factor, or a factor $(a+b)$ of which it is a summand.

0.10 Theorems on Multiplication and Addition

We can now establish two theorems with regard to the combination of inclusions and equalities by addition and multiplication:

Th. 1

$$(a < b) < (ac < bc), \qquad (a < b) < (a + c < b + c).$$

Demonstration:

1 (Simpl.) $\quad ac < c,$

(Syll.) $\quad (ac < a)(a < b) < (ac < b),$

(Comp.) $\quad (ac < b)(ac < c) < (ac < bc).$

2 (Simpl.) $\quad c < b + c,$

(Syll.) $\quad (a < b)(b < b + c) < (a < b + c).$

(Comp.) $\quad (a < b + c)(a < b + c) < (a + c < b + c).$

This theorem may be easily extended to the case of equalities:

$$(a = b) < (ac = bc), \qquad (a = b) < (a + c = b + c).$$

Th. 2

$$(a < b)(c < d) < (ac < bd),$$
$$(a < b)(c < d) < (a + c < b + d).$$

Demonstration:

1 (Syll.) $(ac < a)(a < b) < (ac < b),$

(Syll.) $(ac < c)(c < d) < (ac < a),$

(Comp.) $(ac < b)(ac < d) < (ac < bd).$

2 (Syll.) $(a < b)(b < b + d) < (a < b + d),$

(Syll.) $(c < d)(d < b + d) < (c < b + d),$

(Comp.) $(a < b + d)(c < b + d) < (a + c < b + d).$

This theorem may easily be extended to the case in which one of the two inclusions is replaced by an equality:

$$(a = b)(c < d) < (ac < bd),$$
$$(a = b)(c < d) < (a + c < b + d).$$

When both are replaced by equalities the result is an equality:

$$(a = b)(c = d) < (ac = bd),$$
$$(a = b)(c = d) < (a + c = b + d).$$

To sum up, two or more inclusions or equalities can be added or multiplied together member by member; the result will not be an equality unless all the propositions combined are equalities.

0.11 The First Formula for Transforming Inclusions into Equalities

We can now demonstrate an important formula by which an inclusion may be transformed into an equality, or *vice versa*:

$$(a < b) = (a = ab) \qquad (a < b) = (a + b = b)$$

Demonstration:

1. $(a < b) < (a = ab), \qquad (a < b) < (a + b = b).$

For

(Comp.) $$(a < a)(a < b) < (a < ab),$$
$$(a < b)(b < b) < (a + b < b).$$

On the other hand, we have

(Simpl.) $$ab < a, b < a + b,$$
(Def. =) $$(a < ab)(ab < a) = (a = ab)$$
$$(a + b < b)(b < a + b) = (a + b = b).$$

2. $$(a = ab) < (a < b), \qquad (a + b = b) < (a < b).$$

For

$$(a - ab)(ab < b) < (a < b),$$
$$(a < a + b)(a + b = b) < (a < b).$$

Remark.—If we take the relation of equality as a primitive idea (one not defined) we shall be able to define the relation of inclusion by means of one of the two preceding formulas.[20] We shall then be able to demonstrate the principle of the syllogism.[21]

From the preceding formulas may be derived an interesting result:

$$(a = b) = (ab = a + b).$$

For

1. $$(a = b) = (a < b)(b < a),$$
$$(a < b) = (a = ab), (b < a) = (a + b = a),$$
(Syll.) $$(a = ab)(a + b = a) < (ab = a + b).$$

2. $$(ab = a + b) < (a + b < ab),$$
(Comp.) $$(a + b < ab) = (a < ab)(b < ab),$$
$$(a < ab)(ab < a) = (a = ab) = (a < b),$$
$$(b < ab)(ab < b) = (b = ab) = (b < a),$$

Hence
$$(ab = a + b) < (a < b)(b < a) = (a = b).$$

[20]See HUNTINGTON, *op. cit.*, §??.

[21]This can be demonstrated as follows: By definition we have $(a < b) = (a = ab)$, and $(b < c) = (b = bc)$. If in the first equality we substitute for b its value derived from the second equality, then $a = abc$. Substitute for a its equivalent ab, then $ab = abc$. This equality is equivalent to the inclusion, $ab < c$. Conversely substitute a for ab; whence we have $a < c$. Q.E.D.

0.12 The Distributive Law

The principles previously stated make it possible to demonstrate the *converse distributive law*, both of multiplication with respect to addition, and of addition with respect to multiplication,

$$ac + bc < (a + b)c, \qquad ab + c < (a + c)(b + c).$$

Demonstration:

$$(a < a + b) < [ac < (a + b)c],$$
$$(b < a + b) < [bc < (a + b)c];$$

whence, by composition,

$$[ac < (a + b)c][bc < (a + b)c] < [ac + bc < (a + b)c]$$

2.

$$(ab < a) < (ab + c < a + c),$$
$$(ab < b) < (ab + c < b + c),$$

whence, by composition,

$$(ab + c < a + c)(ab + c < b + c) < [ab + c < (a + c)(b + c)].$$

But these principles are not sufficient to demonstrate the *direct distributive law*

$$(a + b)c < ac + bc, \qquad (a + c)(b + c) < ab + c,$$

and we are obliged to postulate one of these formulas or some simpler one from which they can be derived. For greater convenience we shall postulate the formula

Ax. 5

$$(a + b)c < ac + bc.$$

This, combined with the converse formula, produces the equality

$$(a + b)c = ac + bc$$

which we shall call briefly the *distributive law.*

From this may be directly deduced the formula

$$(a + b)(c + d) = ac + bc + ad + bd,$$

and consequently the second formula of the distributive law,

$$(a + c)(b + c) = ab + c.$$

For

$$(a+c)(b+c) = ab + ac + bc + c,$$

and, by the law of absorption,

$$ac + bc + c = c.$$

This second formula implies the inclusion cited above,

$$(a+c)(b+c) < ab + c,$$

which thus is shown to be proved.

Corollary.—We have the equality

$$ab + ac + bc = (a+b)(a+c)(b+c),$$

for

$$(a+b)(a+c)(b+c) = (a+bc)(b+c) = ab + ac + bc.$$

It will be noted that the two members of this equality differ only in having the signs of multiplication and addition transposed (compare §0.14).

0.13 Definition of 0 and 1

We shall now define and introduce into the logical calculus two special terms which we shall designate by 0 and by 1, because of some formal analogies that they present with the zero and unity of arithmetic. These two terms are formally defined by the two following principles which affirm or postulate their existence.

Ax. 6 There is a term 0 such that whatever value may be given to the term x, we have

$$0 < x.$$

Ax. 7 There is a term 1 such that whatever value may be given to the term x, we have

$$x < 1.$$

It may be shown that each of the terms thus defined is unique; that is to say, if a second term possesses the same property it is equal to (identical with) the first.

The two interpretations of these terms give rise to paradoxes which we shall not stop to elucidate here, but which will be justified by the conclusions of the theory.[22]

C. I.: 0 denotes the class contained in every class; hence it is the "null" or "void" class which contains no element (Nothing or Naught), 1 denotes the class which contains all classes; hence it is the totality of the elements which are

[22] Compare the author's *Manuel de Logistique*, Chap. I., §8, Paris, 1905 [This work, however, did not appear].

contained within it. It is called, after BOOLE, the "universe of discourse" or simply the "whole".

P. I.: 0 denotes the proposition which implies every proposition; it is the "false" or the "absurd", for it implies notably all pairs of contradictory propositions, 1 denotes the proposition which is implied in every proposition; it is the "true", for the false may imply the true whereas the true can imply only the true.

By definition we have the following inclusions

$$0 < 0, \quad 0 < 1, \quad 1 < 1,$$

the first and last of which, moreover, result from the principle of identity. It is important to bear the second in mind.

C. I.: The null class is contained in the *whole.*[23]

P. I.: The false implies the true.

By the definitions of 0 and 1 we have the equivalences

$$(a < 0) = (a = 0), \quad (1 < a) = (a = 1),$$

since we have

$$0 < a, \quad a < 1$$

whatever the value of a.

Consequently the principle of composition gives rise to the two following corollaries:

$$(a = 0)(b = 0) = (a + b = 0),$$
$$(a = 1)(b = 1) = (ab = 1).$$

Thus we can combine two equalities having 0 for a second member by adding their first members, and two equalities having 1 for a second member by multiplying their first members.

Conversely, to say that a sum is "null" [zero] is to say that each of the summands is null; to say that a product is equal to 1 is to say that each of its factors is equal to 1.

Thus we have

$$(a + b = 0) < (a = 0),$$
$$(ab = 1) < (a = 1),$$

and more generally (by the principle of the syllogism)

$$(a < b)(b = 0) < (a = 0),$$
$$(a < b)(a = 1) < (b = 1).$$

It will be noted that we can not conclude from these the equalities $ab = 0$ and $a + b = 1$. And indeed in the conceptual interpretation the first equality

[23] The rendering "Nothing is everything" must be avoided.

denotes that the part common to the classes a and b is null; it by no means follows that either one or the other of these classes is null. The second denotes that these two classes combined form the whole; it by no means follows that either one or the other is equal to the whole.

The following formulas comprising the rules for the calculus of 0 and 1, can be demonstrated:

$$a \times 0 = 0, \quad a + 1 = 1,$$
$$a + 0 = a, \quad a \times 1 = a.$$

For

$$(0 < a) = (0 = 0 \times a) = (a + 0 = a),$$
$$(a < 1) = (a = a \times 1) = (a + 1 = 1).$$

Accordingly it does not change a term to add 0 to it or to multiply it by 1. We express this fact by saying that 0 is the *modulus* of addition and 1 the *modulus* of multiplication.

On the other hand, the product of any term whatever by 0 is 0 and the sum of any term whatever with 1 is 1.

These formulas justify the following interpretation of the two terms:

C. I.: The part common to any class whatever and to the null class is the null class; the sum of any class whatever and of the whole is the whole. The sum of the null class and of any class whatever is equal to the latter; the part common to the whole and any class whatever is equal to the latter.

P. I.: The simultaneous affirmation of any proposition whatever and of a false proposition is equivalent to the latter (*i.e.*, it is false); while their alternative affirmation is equal to the former. The simultaneous affirmation of any proposition whatever and of a true proposition is equivalent to the former; while their alternative affirmation is equivalent to the latter (*i.e.*, it is true).

Remark.—If we accept the four preceding formulas as axioms, because of the proof afforded by the double interpretation, we may deduce from them the paradoxical formulas

$$0 < x, \text{ and } x < 1,$$

by means of the equivalences established above,

$$(a - ab) = (a < b) = (a + b = b).$$

0.14 The Law of Duality

We have proved that a perfect symmetry exists between the formulas relating to multiplication and those relating to addition. We can pass from one class to the other by interchanging the signs of addition and multiplication, on condition that we also interchange the terms 0 and 1 and reverse the meaning of the sign $<$ (or transpose the two members of an inclusion). This symmetry, or *duality* as it is called, which exists in principles and definitions, must also exist in all the formulas deduced from them as long as no principle or definition is introduced which would overthrow them. Hence a true formula may be deduced from another true formula by transforming it by the principle of duality; that is, by following the rule given above. In its application the *law of duality* makes it possible to replace two demonstrations by one. It is well to note that this law is derived from the definitions of addition and multiplication (the formulas for which are reciprocal by duality) and not, as is often thought[24], from the laws of negation which have not yet been stated. We shall see that these laws possess the same property and consequently preserve the duality, but they do not originate it; and duality would exist even if the idea of negation were not introduced. For instance, the equality (§0.12)

$$ab + ac + bc = (a + b)(a + c)(b + c)$$

is its own reciprocal by duality, for its two members are transformed into each other by duality.

It is worth remarking that the law of duality is only applicable to primary propositions. We call [after BOOLE] those propositions *primary* which contain but one copula ($<$ or $=$). We call those propositions *secondary* of which both members (connected by the copula $<$ or $=$) are primary propositions, and so on. For instance, the principle of identity and the principle of simplification are primary propositions, while the principle of the syllogism and the principle of composition are secondary propositions.

0.15 Definition of Negation

The introduction of the terms 0 and 1 makes it possible for us to define *negation*. This is a "uni-nary" operation which transforms a single term into another term called its *negative*.[25] The negative of a is called not-a and is written a'.[26] Its formal definition implies the following postulate of existence[27]:

[24] BOOLE thus derives it (*Laws of Thought*, London 1854, Chap. III, Prop. IV).

[25][In French] the same word *negation* denotes both the operation and its result, which becomes equivocal. The result ought to be denoted by another word, like [the English] "negative". Some authors say, "supplementary" or "supplement", [*e.g.* BOOLE and HUNTINGTON], Classical logic makes use of the term "contradictory" especially for propositions.

[26]We adopt here the notation of MACCOLL; SCHRÖDER indicates not-a by a_1 which prevents the use of indices and obliges us to express them as exponents. The notation a' has the advantage of excluding neither indices nor exponents. The notation $\bar{a}$ employed by many authors is inconvenient for typographical reasons. When the negative affects a proposition written in an explicit form (with a copula) it is applied to the copula $<$ or $=$) by a vertical bar ($\not<$) or $\neq$). The accent can be considered as the indication of a vertical bar applied to letters.

[27] BOOLE follows Aristotle in usually calling the law of duality the principle of contradiction "which affirms that it is impossible for any being to possess a quality and at the same time not to possess it". He writes it in the form of an equation of the second degree, $x - x^2 = 0$, or $x(1 - x) = 0$ in which $1 - x$ expresses the universe less x, or not x. Thus he regards the law of duality as derived from negation as stated in note 24 above.

Ax. 8 Whatever the term a may be, there is also a term a' such that we have at the same time

$$aa' = 0, \quad a + a' = 1.$$

It can be proved by means of the following *lemma* that if a term so defined exists it is unique:

If at the same time

$$ac = bc, \quad a + c = b + c,$$

then

$$a = b.$$

Demonstration.—Multiplying both members of the second premise by a, we have

$$a + ac = ab + ac.$$

Multiplying both members by b,

$$ab + bc = b + bc.$$

By the first premise,

$$ab + ac = ab + bc.$$

Hence

$$a + ac = b + bc,$$

which by the law of absorption may be reduced to

$$a = b.$$

Remark.—This demonstration rests upon the direct distributive law. This law cannot, then, be demonstrated by means of negation, at least in the system of principles which we are adopting, without reasoning in a circle.

This lemma being established, let us suppose that the same term a has two negatives; in other words, let a'_1 and a'_2 be two terms each of which by itself satisfies the conditions of the definition. We will prove that they are equal. Since, by hypothesis,

$$aa'_1 = 0, \quad a + a'_1 = 1,$$
$$aa'_2 = 0, \quad a + a'_2 = 1,$$

we have

$$aa'_1 = aa'_2, \quad a + a'_1 = a + a'_2;$$

whence we conclude, by the preceding lemma, that

$$a'_1 = a'_2.$$

We can now speak of *the* negative of a term as of a unique and well-defined term.

The *uniformity* of the operation of negation may be expressed in the following manner:

If $a = b$, then also $a' = b'$. By this proposition, both members of an equality in the logical calculus may be "denied".

0.16 The Principles of Contradiction and of Excluded Middle

By definition, a term and its negative verify the two formulas

$$aa' = 0, \quad a + a' = 1,$$

which represent respectively the *principle of contradiction* and the *principle of excluded middle.*[28]

C. I.: 1. The classes a and a' have nothing in common; in other words, no element can be at the same time both a and not-a.

2. The classes a and a' combined form the whole; in other words, every element is either a or not-a.

P. I.: 1. The simultaneous affirmation of the propositions a and not-a is false; in other words, these two propositions cannot both be true at the same time.

2. The alternative affirmation of the propositions a and not-a is true; in other words, one of these two propositions must be true.

Two propositions are said to be *contradictory* when one is the negative of the other; they cannot both be true or false at the same time. If one is true the other is false; if one is false the other is true.

This is in agreement with the fact that the terms 0 and 1 are the negatives of each other; thus we have

$$0 \times 1 = 0, \quad 0 + 1 = 1.$$

Generally speaking, we say that two terms are *contradictory* when one is the negative of the other.

0.17 Law of Double Negation

Moreover this reciprocity is general: if a term b is the negative of the term a, then the term a is the negative of the term b. These two statements are expressed by the same formulas

$$ab = 0, \quad a + b = 1,$$

and, while they unequivocally determine b in terms of a, they likewise determine a in terms of b. This is due to the symmetry of these relations, that is to say, to the commutativity of multiplication and addition. This reciprocity is expressed by the *law of double negation*

$$(a')' = a,$$

[28]As Mrs. Ladd-Franklin has truly remarked (Baldwin, *Dictionary of Philosophy and Psychology*, article "Laws of Thought"), the principle of *contradiction* is not sufficient to define *contradictories*; the principle of excluded middle must be added which equally deserves the name of principle of contradiction. This is why Mrs. Ladd-Franklin proposes to call them respectively the *principle of exclusion* and the *principle of exhaustion*, inasmuch as, according to the first, two contradictory terms are *exclusive* (the one of the other); and, according to the second, they are *exhaustive* (of the universe of discourse).

which may be formally proved as follows: a' being by hypothesis the negative of a, we have

$$aa' = 0, \quad a + a' = 1.$$

On the other hand, let a'' be the negative of a'; we have, in the same way,

$$a'a'' = 0, \quad a' + a'' = 1.$$

But, by the preceding lemma, these four equalities involve the equality

$$a = a''.$$

Q. E. D.

This law may be expressed in the following manner:

If $b = a'$, we have $a = b'$, and conversely, by symmetry.

This proposition makes it possible, in calculations, to transpose the negative from one member of an equality to the other.

The law of double negation makes it possible to conclude the equality of two terms from the equality of their negatives (if $a' = b'$ then $a = b$), and therefore to cancel the negation of both members of an equality.

From the characteristic formulas of negation together with the fundamental properties of 0 and 1, it results that every product which contains two contradictory factors is null, and that every sum which contains two contradictory summands is equal to 1.

In particular, we have the following formulas:

$$a = ab + ab', \quad a = (a + b)(a + b'),$$

which may be demonstrated as follows by means of the distributive law:

$$a = a \times 1 = a(b + b') = ab + ab',$$
$$a = a + 0 = a + bb' = (a + b)(a + b').$$

These formulas indicate the principle of the method of development which we shall explain in detail later (§§0.21 sqq.)

0.18 Second Formulas for Transforming Inclusions into Equalities

We can now establish two very important equivalences between inclusions and equalities:

$$(a < b) = (ab' = 0), \quad (a < b) = (a' + b = 1).$$

Demonstration.—1. If we multiply the two members of the inclusion $a < b$ by b' we have

$$(ab' < bb') = (ab' < 0) = (ab' = 0).$$

2. Again, we know that

$$a = ab + ab'.$$

Now if $ab' = 0$,

$$a = ab + 0 = ab.$$

On the other hand: 1. Add a' to each of the two members of the inclusion $a < b$; we have

$$(a' + a < a' + b) = (1 < a' + b) = a' + b = 1).$$

2. We know that

$$b = (a + b)(a' + b).$$

Now, if $a' + b = 1$,

$$b = (a + b) \times 1 = a + b.$$

By the preceding formulas, an inclusion can be transformed at will into an equality whose second member is either 0 or 1. Any equality may also be transformed into an equality of this form by means of the following formulas:

$$(a = b) = (ab' + a'b = 0), \quad (a = b) = [(a + b')(a' + b) = 1].$$

Demonstration:

$$(a = b) = (a < b)(b < a) = (ab' = 0)(a'b = 0) = (ab' + a'b = 0),$$
$$(a = b) = (a < b)(b < a) = (a' + b = 1)(a + b' = 1) = [(a' + b')(a' + b) = 1].$$

Again, we have the two formulas

$$(a = b) = [(a + b)(a' + b') = 0], \quad (a = b) = (ab + a'b' = 1),$$

which can be deduced from the preceding formulas by performing the indicated multiplications (or the indicated additions) by means of the distributive law.

0.19 The Law of Contraposition

We are now able to demonstrate the *law of contraposition*,

$$(a < b) = (b' < a').$$

Demonstration.—By the preceding formulas, we have

$$(a < b) = (ab = 0) = (b' < a').$$

Again, the law of contraposition may take the form

$$(a < b') = (b < a'),$$

which presupposes the law of double negation. It may be expressed verbally as follows: "Two members of an inclusion may be interchanged on condition that both are denied".

C. I.: "If all a is b, then all not-b is not-a, and conversely".

P. I.: "If a implies b, not-b implies not-a and conversely"; in other words, "If a is true b is true", is equivalent to saying, "If b is false, a is false".

This equivalence is the principle of the *reductio ad absurdum* (see hypothetical arguments, *modus tollens*, §0.58).

0.20 Postulate of Existence

One final axiom may be formulated here, which we will call the *postulate of existence*:

Ax. 9

$$1 \not< 0$$

whence may be also deduced $1 \neq 0$.

In the conceptual interpretation (C. I.) this axiom means that the universe of discourse is not null, that is to say, that it contains some elements, at least one. If it contains but one, there are only two classes possible, 1 and 0. But even then they would be distinct, and the preceding axiom would be verified.

In the propositional interpretation (P. I.) this axiom signifies that the true and false are distinct; in this case, it bears the mark of evidence and necessity. The contrary proposition, $1 = 0$, is, consequently, the type of *absurdity* (of the formally false proposition) while the propositions $0 = 0$, and $1 = 1$ are types of *identity* (of the formally true proposition). Accordingly we put

$$(1 = 0) = 0, \quad (0 = 0) = (l = 1) = 1.$$

More generally, every equality of the form

$$x = x$$

is equivalent to one of the identity-types; for, if we reduce this equality so that its second member will be 0 or 1, we find

$$(xx' + xx' = 0) = (0 = 0), \quad (xx + x'x' = 1) = (1 = 1).$$

On the other hand, every equality of the form

$$x = x'$$

is equivalent to the absurdity-type, for we find by the same process,

$$(xx + x'x' = 0) = (1 = 0), \quad (xx' + xx' = 1) = (0 = 1).$$

0.21 The Development of 0 and of 1

Hitherto we have met only such formulas as directly express customary modes of reasoning and consequently offer direct evidence.

We shall now expound theories and methods which depart from the usual modes of thought and which constitute more particularly the algebra of logic in so far as it is a formal and, so to speak, automatic method of an absolute universality and an infallible certainty, replacing reasoning by calculation.

The fundamental process of this method is *development.* Given the terms $a, b, c \ldots$ (to any finite number), we can develop 0 or 1 with respect to these terms (and their negatives) by the following formulas derived from the distributive law:

$$\begin{aligned}
&0 = aa', \\
&0 = aa' + bb' = (a+b)(a+b')(a'+b)(a'+b'), \\
&0 = aa' + bb' + cc' = (a+b+c)(a+b+c')(a+b'+c) \\
&\qquad\qquad\qquad\qquad \times (a+b'+c')(a'+b+c) \\
&\qquad\qquad\qquad\qquad \times (a'+b+c')(a'+b'+c)(a'+b'+c'); \\
&1 = a + a', \\
&1 = (a+a')(b+b') = ab + ab' + a'b + a'b', \\
&1 = (a+a')(b+b')(c+c') = abc + abc' + ab'c + ab'c' \\
&\qquad\qquad\qquad\qquad + a'bc + a'bc' + a'b'c + a'b'c';
\end{aligned}$$

and so on. In general, for any number n of simple terms; 0 will be developed in a product containing 2^n factors, and 1 in a sum containing 2^n summands. The factors of zero comprise all possible additive combinations, and the summands of 1 all possible multiplicative combinations of the n given terms and their negatives, each combination comprising n different terms and never containing a term and its negative at the same time.

The summands of the development of 1 are what BOOLE called the *constituents* (of the universe of discourse). We may equally well call them, with PORETSKY, [29] the *minima* of discourse, because they are the smallest classes into which the universe of discourse is divided with reference to the n given terms. In the same way we shall call the factors of the development of 0 the *maxima* of discourse, because they are the largest classes that can be determined in the universe of discourse by means of the n given terms.

0.22 Properties of the Constituents

The constituents or minima of discourse possess two properties characteristic of contradictory terms (of which they are a generalization); they are *mutually exclusive*, *i.e.*, the product of any two of them is 0; and they are *collectively exhaustive*, *i.e.*, the sum of all "exhausts" the universe of discourse. The latter

[29] See the Bibliography, page xiv.

property is evident from the preceding formulas. The other results from the fact that any two constituents differ at least in the "sign" of one of the terms which serve as factors, *i.e.*, one contains this term as a factor and the other the negative of this term. This is enough, as we know, to ensure that their product be null.

The maxima of discourse possess analogous and correlative properties; their combined product is equal to 0, as we have seen; and the sum of any two of them is equal to 1, inasmuch as they differ in the sign of at least one of the terms which enter into them as summands.

For the sake of simplicity, we shall confine ourselves, with BOOLE and SCHRÖDER, to the study of the constituents or minima of discourse, *i.e.*, the developments of 1. We shall leave to the reader the task of finding and demonstrating the corresponding theorems which concern the maxima of discourse or the developments of 0.

0.23 Logical Functions

We shall call a *logical function* any term whose expression is complex, that is, formed of letters which denote simple terms together with the signs of the three logical operations.[30]

A logical function may be considered as a function of all the terms of discourse, or only of some of them which may be regarded as unknown or variable and which in this case are denoted by the letters x, y, z. We shall represent a function of the variables or unknown quantities, x, y, z, by the symbol $f(x, y, z)$ or by other analogous symbols, as in ordinary algebra. Once for all, a logical function may be considered as a function of any term of the universe of discourse, whether or not the term appears in the explicit expression of the function.

0.24 The Law of Development

This being established, we shall proceed to develop a function $f(x)$ with respect to x. Suppose the problem solved, and let

$$ax + bx'$$

be the development sought. By hypothesis we have the equality

$$f(x) = ax + bx'$$

for all possible values of x. Make $x = 1$ and consequently $x' = 0$. We have

$$f(1) = a.$$

[30] In this algebra the logical function is analogous to the *integral function* of ordinary algebra, except that it has no powers beyond the first.

Then put $x = 0$ and $x' = 1$; we have

$$f(0) = b.$$

These two equalities determine the coefficients a and b of the development which may then be written as follows:

$$f(x) = f(1)x + f(0)x',$$

in which $f(1)$, $f(0)$ represent the value of the function $f(x)$ when we let $x = 1$ and $x = 0$ respectively.

Corollary.—Multiplying both members of the preceding equalities by x and x' in turn, we have the following pairs of equalities (MacColl):

$$xf(x) = ax \qquad x'f(x) = bx'$$
$$xf(x) = xf(1), \quad x'f(x) = x'f(0).$$

Now let a function of two (or more) variables be developed with respect to the two variables x and y. Developing $f(x, y)$ first with respect to x, we find

$$f(x, y) = f(1, y)x + f(0, y)x'.$$

Then, developing the second member with respect to y, we have

$$f(x, y) = f(1, 1)xy + f(1, 0)xy' + f(0, 1)x'y + f(0, 0)x'y'$$

This result is symmetrical with respect to the two variables, and therefore independent of the order in which the developments with respect to each of them are performed.

In the same way we can obtain progressively the development of a function of $3, 4, \ldots\ldots,$ variables.

The general law of these developments is as follows:

To develop a function with respect to n variables, form all the constituents of these n variables and multiply each of them by the value assumed by the function when each of the simple factors of the corresponding constituent is equated to 1 (which is the same thing as equating to 0 those factors whose negatives appear in the constituent).

When a variable with respect to which the development is made, y for instance, does not appear explicitly in the function ($f(x)$ for instance), we have, according to the general law,

$$f(x) = f(x)y + f(x)y'.$$

In particular, if a is a constant term, independent of the variables with respect to which the development is made, we have for its successive developments,

$$a = ax + ax',$$
$$a = axy + axy' + ax'y + ax'y',$$
$$a = axyz + axyz' + axy'z + axy'z' + ax'yz + ax'yz' + ax'y'z + ax'y'z'.$$[31]

and so on. Moreover these formulas may be directly obtained by multiplying by a both members of each development of 1.

Cor. 1. We have the equivalence

$$(a + x')(b + x) = ax + bx + ab = ax + bx'.$$

For, if we develop with respect to x, we have

$$ax + bx' + abx + abx' = (a + ab)x + (b + ab)x' = ax + bx'.$$

Cor. 2. We have the equivalence

$$ax + bx' + c = (a + c)x + (b + c)x'.$$

For if we develop the term c with respect to x, we find

$$ax + bx' + cx + cx' = (a + c)x + (b + c)x'.$$

Thus, when a function contains terms (whose sum is represented by c) independent of x, we can always reduce it to the developed form $ax+bx'$ by adding c to the coefficients of both x and x'. Therefore we can always consider a function to be reduced to this form.

In practice, we perform the development by multiplying each term which does not contain a certain letter (x for instance) by $(x + x')$ and by developing the product according to the distributive law. Then, when desired, like terms may be reduced to a single term.

0.25 The Formulas of De Morgan

In any development of 1, the sum of a certain number of constituents is the negative of the sum of all the others.

For, by hypothesis, the sum of these two sums is equal to 1, and their product is equal to 0, since the product of two different constituents is zero.

From this proposition may be deduced the formulas of De Morgan:

$$(a + b)' = a'b', \quad (ab)' = a' + b'.$$

Demonstration.—Let us develop the sum $(a + b)$:

$$a + b = ab + ab' + ab + a'b = ab + ab' + a'b.$$

Now the development of 1 with respect to a and b contains the three terms of this development plus a fourth term $a'b'$. This fourth term, therefore, is the negative of the sum of the other three.

We can demonstrate the second formula either by a correlative argument (*i.e.*, considering the development of 0 by factors) or by observing that the development of $(a' + b')$,

$$a'b + ab' + a'b',$$

[31]These formulas express the method of classification by dichotomy.

differs from the development of 1 only by the summand ab.

How DE MORGAN'S formulas may be generalized is now clear; for instance we have for a sum of three terms,

$$a + b + c = abc + abc' + ab'c + ab'c' + a'bc + a'bc' + a'b'c.$$

This development differs from the development of 1 only by the term $a'b'c'$. Thus we can demonstrate the formulas

$$(a + b + c)' = a'b'c', \quad (abc)' = a' + b' + c',$$

which are generalizations of DE MORGAN'S formulas.

The formulas of DE MORGAN are in very frequent use in calculation, for they make it possible to perform the negation of a sum or a product by transferring the negation to the simple terms: the negative of a sum is the product of the negatives of its summands; the negative of a product is the sum of the negatives of its factors.

These formulas, again, make it possible to pass from a primary proposition to its correlative proposition by duality, and to demonstrate their equivalence. For this purpose it is only necessary to apply the law of contraposition to the given proposition, and then to perform the negation of both members.

Example:

$$ab + ac + bc = (a + b)(a + c)(b + c).$$

Demonstration:

$$\begin{aligned} (ab + ac + bc)' &= [(a + b)(a + c)(b + c)], \\ (ab)'(ac)'(bc)' &= (a + b)' + (a + c)' + (b + c)', \\ (a' + b')(a' + c')(b' + c') &= a'b' + a'c' + b'c'. \end{aligned}$$

Since the simple terms, a, b, c, may be any terms, we may suppress the sign of negation by which they are affected, and obtain the given formula.

Thus DE MORGAN'S formulas furnish a means by which to find or to demonstrate the formula correlative to another; but, as we have said above (§0.14), they are not the basis of this correlation.

0.26 Disjunctive Sums

By means of development we can transform any sum into a *disjunctive* sum, *i.e.*, one in which each product of its summands taken two by two is zero. For, let $(a+b+c)$ be a sum of which we do not know whether or not the three terms are disjunctive; let us assume that they are not. Developing, we have:

$$a + b + c = abc + abc' + ab'c + ab'c' + a'bc + a'bc' + a'b'c.$$

Now, the first four terms of this development constitute the development of a with respect to b and c; the two following are the development of $a'b$ with respect to c. The above sum, therefore, reduces to

$$a + a'b + a'b'c,$$

and the terms of this sum are disjunctive like those of the preceding, as may be verified. This process is general and, moreover, obvious. To enumerate without repetition all the a's, all the b's, and all the c's, etc., it is clearly sufficient to enumerate all the a's, then all the b's which are not a's, and then all the c's which are neither a's nor b's, and so on.

It will be noted that the expression thus obtained is not symmetrical, since it depends on the order assigned to the original summands. Thus the same sum may be written:

$$b + ab' + a'b'c, \quad c + ac' + a'bc', \ldots.$$

Conversely, in order to simplify the expression of a sum, we may suppress as factors in each of the summands (arranged in any suitable order) the negatives of each preceding summand. Thus, we may find a symmetrical expression for a sum. For instance,

$$a + a'b = b + ab' = a + b.$$

0.27 Properties of Developed Functions

The practical utility of the process of development in the algebra of logic lies in the fact that developed functions possess the following property:

The sum or the product of two functions developed with respect to the same letters is obtained simply by finding the sum or the product of their coefficients. The negative of a developed function is obtained simply by replacing the coefficients of its development by their negatives.

We shall now demonstrate these propositions in the case of two variables; this demonstration will of course be of universal application.

Let the developed functions be

$$a_1xy + b_1xy' + c_1x'y + d_1x'y',$$
$$a_2xy + b_2xy' + c_2x'y + d_2x'y'.$$

1. I say that their sum is

$$(a_1 + a_2)xy + (b_1 + b_2)xy' + (c_1 + c_2)x'y + (d_1 + d_2)x'y'.$$

This result is derived directly from the distributive law.

2. I say that their product is

$$a_1a_2xy + b_1b_2xy' + c_1c_2x'y + d_1d_2x'y',$$

for if we find their product according to the general rule (by applying the distributive law), the products of two terms of different constituents will be zero; therefore there will remain only the products of the terms of the same constituent, and, as (by the law of tautology) the product of this constituent multiplied by itself is equal to itself, it is only necessary to obtain the product of the coefficients.

3. Finally, I say that the negative of

$$axy + bxy' + cx'y + dx'y'$$

is

$$a'xy + b'xy' + c'x'y + d'x'y'.$$

In order to verify this statement, it is sufficient to prove that the product of these two functions is zero and that their sum is equal to 1. Thus

$$\begin{aligned}
&(axy + bxy' + cx'y + dx'y')(a'xy + b'xy' + c'x'y + d'x'y') \\
&\quad = (aa'xy + bb'xy' + cc'x'y + dd'x'y') \\
&\quad = (0 \cdot xy + 0 \cdot xy' + 0 \cdot x'y + 0 \cdot x'y') = 0 \\
&(axy + bxy' + cx'y + dx'y') + (a'xy + b'xy' + c'x'y + d'x'y') \\
&\quad = [(a + a')xy + (b + b')xy' + (c + c')x'y + (d + d')x'y'] \\
&\quad = (1xy + 1xy' + 1x'y + 1x'y') = 1.
\end{aligned}$$

Special Case.—We have the equalities:

$$\begin{aligned}
(ab + a'b')' &= ab' + a'b, \\
(ab' + a'b')' &= ab + a'b',
\end{aligned}$$

which may easily be demonstrated in many ways; for instance, by observing that the two sums $(ab + a'b')$ and $(ab' + a'b)$ combined form the development of 1; or again by *performing* the negation $(ab + a'b')'$ by means of De Morgan's formulas (§0.25).

From these equalities we can deduce the following equality:

$$(ab' + ab = 0) = (ab + a'b' = 1),$$

which result might also have been obtained in another way by observing that (§0.18)

$$(a = b) = (ab' + a'b = 0) = [(a + b')(a' + b) = 1],$$

and by performing the multiplication indicated in the last equality.

Theorem.—*We have the following equivalences:*[32]

$$(a = bc' + b'c) = (b = ac' + a'c) = (c = ab' + a'b).$$

For, reducing the first of these equalities so that its second member will be 0,

$$\begin{aligned}
a(bc + b'c') + a'(bc' + b'c) &= 0, \\
abc + ab'c' + a'bc' + a'b'c &= 0.
\end{aligned}$$

Now it is clear that the first member of this equality is symmetrical with respect to the three terms a, b, c. We may therefore conclude that, if the two

[32] W. Stanley Jevons, *Pure Logic*, 1864, p. 61.

other equalities which differ from the first only in the permutation of these three letters be similarly transformed, the same result will be obtained, which proves the proposed equivalence.

Corollary.—If we have at the same time the three inclusions:

$$a < bc' + b'c, \quad b < ac' + a'c, \quad c < ab' + a'b.$$

we have also the converse inclusion, an therefore the corresponding equalities

$$a = bc' + b'c, \quad b = ac' + a'c, \quad c = ab' + a'b.$$

For if we transform the given inclusions into equalities, we shall have

$$abc + ab'c' = 0, \quad abc + a'bc' = 0, \quad abc + a'b'c = 0,$$

whence, by combining them into a single equality,

$$abc + ab'c' + a'bc' + a'b'c = 0.$$

Now this equality, as we see, is equivalent to any one of the three equalities to be demonstrated.

0.28 The Limits of a Function

A term x is said to be *comprised* between two given terms, a and b, when it contains one and is contained in the other; that is to say, if we have, for instance,

$$a < x, \quad x < b,$$

which we may write more briefly as

$$a < x < b.$$

Such a formula is called a *double inclusion.* When the term x is variable and always comprised between two constant terms a and b, these terms are called the *limits* of x. The first (contained in x) is called *inferior limit*; the second (which contains x) is called the *superior limit.*

THEOREM. —*A developed function is comprised between the sum and the product of its coefficients.*

We shall first demonstrate this theorem for a function of one variable,

$$ax + bx'.$$

We have, on the one hand,

$$(ab < a) < (abx < ax),$$
$$(ab < b) < (abx' < bx').$$

Therefore

$$abx + abx' < ax + bx',$$

or

$$ab < ax + bx'.$$

On the other hand,

$$(a < a + b) < [ax < (a + b)x],$$
$$(b < a + b) < [bx' < (a + b)x'].$$

Therefore

$$ax + bx' < (a + b)(x + x'),$$

or

$$ax + bx' < a + b.$$

To sum up,

$$ab < ax + bx' < a + b.$$

Q. E. D.

Remark 1. This double inclusion may be expressed in the following form:[33]

$$f(b) < f(x) < f(a).$$

For

$$f(a) = aa + ba' = a + b,$$
$$f(b) = ab + bb' = ab.$$

But this form, pertaining as it does to an equation of one unknown quantity, does not appear susceptible of generalization, whereas the other one does so appear, for it is readily seen that the former demonstration is of general application. Whatever the number of variables n (and consequently the number of constituents 2^n) it may be demonstrated in exactly the same manner that the function contains the product of its coefficients and is contained in their sum. Hence the theorem is of general application.

Remark 2.—This theorem assumes that all the constituents appear in the development, consequently those that are wanting must really be present with the coefficient 0. In this case, the product of all the coefficients is evidently 0. Likewise when one coefficient has the value 1, the sum of all the coefficients is equal to 1.

It will be shown later (§0.38) that a function may reach both its limits, and consequently that they are its extreme values. As yet, however, we know only that it is always comprised between them.

[33] Eugen Müller, *Aus der Algebra der Logik*, Art. II.

0.29 Formula of Poretsky.[34]

We have the equivalence

$$(x = ax + bx') = (b < x < a).$$

Demonstration.—First multiplying by x both members of the given equality [which is the first member of the entire secondary equality], we have

$$x = ax,$$

which, as we know, is equivalent to the inclusion

$$x < a.$$

Now multiplying both members by x', we have

$$0 = bx',$$

which, as we know, is equivalent to the inclusion

$$b < x.$$

Summing up, we have

$$(x = ax + bx') < (b < x < a).$$

Conversely,

$$(b < x < a) < (x = ax + bx').$$

For

$$(x < a) = (x = ax),$$
$$(b < x) = (bx' = 0).$$

Adding these two equalities member to member [the second members of the two larger equalities],

$$(x = ax)(o = bx) < (x = ax + bx').$$

Therefore

$$(b < x < a) < (x = ax + bx')$$

and thus the equivalence is proved.

[34] Poretsky, "Sur les méthodes pour résoudre les égalités logiques". (*Bull. de la Soc. phys.-math. de Kazan*, Vol. II, 1884).

0.30 Schröder's Theorem.[35]

The equality

$$ax + bx' = 0$$

signifies that x lies between a' and b.

Demonstration:

$$(ax + bx' = 0) = (ax = 0)(bx' = 0),$$
$$(ax = 0) = (x < a'),$$
$$(bx' = 0) = (b < x).$$

Hence

$$(ax + bx' = 0) = (b < x < a').$$

Comparing this theorem with the formula of PORETSKY, we obtain at once the equality

$$(ax + bx' = 0) = (x = a'x + bx'),$$

which may be directly proved by reducing the formula of PORETSKY to an equality whose second member is 0, thus:

$$(x = a'x + bx') = [x(ax + b'x') + x'(a'x + bx') = 0] = (ax + bx' = 0).$$

If we consider the given equality as an *equation* in which x is the unknown quantity, PORETSKY'S formula will be its solution.

From the double inclusion

$$b < x < a'$$

we conclude, by the principle of the syllogism, that

$$b < a'$$

This is a consequence of the given equality and is independent of the unknown quantity x. It is called the *resultant of the elimination* of x in the given equation. It is equivalent to the equality

$$ab = 0.$$

Therefore we have the implication

$$(ax + bx' = 0) < (ab = 0).$$

Taking this consequence into consideration, the solution may be simplified, for

$$(ab = 0) = (b = a'b).$$

[35] SCHRÖDER, *Operationskreis des Logikkalküls* (1877), Theorem 20.

Therefore

$$\begin{aligned} x &= a'x + bx' = a'x + a'bx' \\ &= a'bx + a'b'x + a'bx' = a'b + a'b'x \\ &= b + a'b'x + b + a'x. \end{aligned}$$

This form of the solution conforms most closely to common sense: since x' contains b and is contained in a', it is natural that x should be equal to the sum of b and a part of a' (that is to say, the part common to a' and x). The solution is generally indeterminate (between the limits a' and b); it is determinate only when the limits are equal,

$$a' = b,$$

for then

$$x = b + a'x = b + bx = b = a'.$$

Then the equation assumes the form

$$(ax + a'x' = 0) = (a' = x)$$

and is equivalent to the double inclusion

$$(a' < x < a') = (x = a').$$

0.31 The Resultant of Elimination

When ab is not zero, the equation is impossible (always false), because it has a false consequence. It is for this reason that SCHRÖDER considers the resultant of the elimination as a *condition* of the equation. But we must not be misled by this equivocal word. The resultant of the elimination of x is not a *cause* of the equation, it is a *consequence* of it; it is not a *sufficient* but a *necessary* condition.

The same conclusion may be reached by observing that ab is the inferior limit of the function $ax+bx'$, and that consequently the function can not vanish unless this limit is 0.

$$(ab < ax + bx')(ax + bx' = 0) < (ab = 0).$$

We can express the resultant of elimination in other equivalent forms; for instance, if we write the equation in the form

$$(a + x')(b + x) = 0,$$

we observe that the resultant

$$ab = 0$$

is obtained simply by dropping the unknown quantity (by suppressing the terms x and x'). Again the equation may be written:

$$a'x + b'x' = 1$$

and the resultant of elimination:

$$a' + b' = 1.$$

Here again it is obtained simply by dropping the unknown quantity.[36]

Remark. If in the equation

$$ax + bx' = 0$$

we substitute for the unknown quantity x its value derived from the equations,

$$x = a'x + bx', \quad x' = ax + b'x',$$

we find

$$(abx + abx' = 0) = (ab = 0),$$

that is to say, the resultant of the elimination of x which, as we have seen, is a consequence of the equation itself. Thus we are assured that the value of x verifies this equation. Therefore we can, with VOIGT, define the solution of an equation as that value which, when substituted for x in the equation, reduces it to the resultant of the elimination of x.

Special Case.—When the equation contains a term independent of x, *i.e.*, when it is of the form

$$ax + bx' + c = 0$$

it is equivalent to

$$(a + c)x + (b + c)x' = 0,$$

and the resultant of elimination is

$$(a + c)(b + c) = ab + c = 0,$$

whence we derive this practical rule: To obtain the resultant of the elimination of x in this case, it is sufficient to equate to zero the product of the coefficients of x and x', and add to them the term independent of x.

[36]This is the method of elimination of Mrs. LADD-FRANKLIN and Mr. MITCHELL, but this rule is deceptive in its apparent simplicity, for it cannot be applied to the same equation when put in either of the forms

$$ax + bx' = 0, \quad (a' + x')(b' + x) = 1.$$

Now, on the other hand, as we shall see (§0.54), for inequalities it may be applied to the forms

$$ax + bx' \neq 0, \quad (a' + x')(b' + x) \neq 1.$$

and not to the equivalent forms

$$(a + x')(b + x) \neq 0, \quad a'x + b'x' \neq 1.$$

Consequently, it has not the mnemonic property attributed to it, for, to use it correctly, it is necessary to recall to which forms it is applicable.

0.32 The Case of Indetermination

Just as the resultant

$$ab = 0$$

corresponds to the case when the equation is possible, so the equality

$$a + b = 0$$

corresponds to the case of ***absolute indetermination.*** For in this case the equation both of whose coefficients are zero $(a = 0)$, $(b = 0)$, is reduced to an identity $(0 = 0)$, and therefore is "identically" verified, whatever the value of x may be; it does not determine the value of x at all, since the double inclusion

$$b < x < a'$$

then becomes

$$0 < x < 1$$

which does not limit in any way the variability of x. In this case we say that the equation is *indeterminate.*

We shall reach the same conclusion if we observe that $(a + b)$ is the superior limit of the function $ax + bx$ and that, if this limit is 0, the function is necessarily zero for all values of x,

$$(ax + bx' < a + b)(a + b = 0) < (ax + bx' = 0).$$

Special Case.—When the equation contains a term independent of x,

$$ax + bx' + c = 0,$$

the condition of absolute indetermination takes the form

$$a + b + c = 0.$$

For

$$ax + bx' + c = (a + c)x + (b + c)x',$$
$$(a + c) + (b + c) = a + b + c = 0.$$

0.33 Sums and Products of Functions

It is desirable at this point to introduce a notation borrowed from mathematics, which is very useful in the algebra of logic. Let $f(x)$ be an expression containing one variable; suppose that the class of all the possible values of x is determined; then the class of all the values which the function $f(x)$ can assume in consequence will also be determined. Their sum will be represented by $\sum_x f(x)$ and their product by $\prod_x f(x)$ This is a new notation and not a new notion, for it is merely the idea of sum and product applied to the values of a function.

When the symbols $\sum$ and $\prod$ are applied to propositions, they assume an interesting significance:

$$\prod_x [f(x) = 0]$$

means that $f(x) = 0$ is true for *every* value of x; and

$$\sum_x [f(x) = 0]$$

that $f(x) = 0$ is true for *some* value of x. For, in order that a product may be equal to 1 (*i.e.*, be true), all its factors must be equal to 1 (*i.e.*, be true); but, in order that a sum may be equal to 1 (*i.e.*, be true), it is sufficient that only one of its summands be equal to 1 (*i.e.*, be true). Thus we have a means of expressing universal and particular propositions when they are applied to variables, especially those in the form: "For every value of x such and such a proposition is true", and "For some value of x, such and such a proposition is true", etc.

For instance, the equivalence

$$(a = b) = (ac = bc)(a + c = b + c)$$

is somewhat paradoxical because the second member contains a term (c) which does not appear in the first. This equivalence is independent of c, so that we can write it as follows, considering c as a variable x

$$\prod_x [(a = b) = (ax = bx)(a + x = b + x)],$$

or, the first member being independent of x,

$$(a = b) = \prod_x [(ax = bx)(a + x = b + x)].$$

In general, when a proposition contains a variable term, great care is necessary to distinguish the case in which it is true for *every* value of the variable, from the case in which it is true only for some value of the variable.[37] This is the purpose that the symbols $\prod$ and $\sum$ serve.

Thus when we say for instance that the equation

$$ax + bx' = 0$$

is possible, we are stating that it can be verified by some value of x; that is to say,

$$\sum_x (ax + bx' = 0),$$

[37]This is the same as the distinction made in mathematics between *identities* and *equations*, except that an equation may not be verified by any value of the variable.

and, since the necessary and sufficient condition for this is that the resultant $(ab = 0)$ is true, we must write

$$\sum_x (ax + bx = 0) = (ab = 0),$$

although we have only the implication

$$(ax + bx = 0) < (ab = 0).$$

On the other hand, the necessary and sufficient condition for the equation to be verified by every value of x is that

$$a + b = 0.$$

Demonstration.—1. The condition is sufficient, for if

$$(a + b = 0) = (a = 0)(b = 0),$$

we obviously have

$$ax + bx' = 0$$

whatever the value of x; that is to say,

$$\prod_x (ax + bx' = 0).$$

2. The condition is necessary, for if

$$\prod_x (ax + bx') = 0,$$

the equation is true, in particular, for the value $x = a$; hence

$$a + b = 0.$$

Therefore the equivalence

$$\prod_x (ax + bx' = 0) = (a + b = 0)$$

is proved.[38] In this instance, the equation reduces to an identity: its first member is "identically" null.

[38] EUGEN MÜLLER, *op. cit.*

0.34 The Expression of an Inclusion by Means of an Indeterminate

The foregoing notation is indispensable in almost every case where variables or indeterminates occur in one member of an equivalence, which are not present in the other. For instance, certain authors predicate the two following equivalences

$$(a < b) = (a = bu) = (a + v = b),$$

in which u, v are two "indeterminates". Now, each of the two equalities has the inclusion $(a < b)$ as its consequence, as we may assure ourselves by eliminating u and v respectively from the following equalities:

1. $$[a(b' + u') + a'bu = 0] = [(ab' + a'b)u + au' = 0].$$

Resultant:

$$[(ab' + a'b)a = 0] = (ab' = 0) = (a < b).$$

2. $$[(a + v)b' + a'bv = 0] = [b'v + (ab' + a'b)v' = 0].$$

Resultant:

$$[b'(ab' + a'b) = 0] = (ab' = 0) + (a < b).$$

But we cannot say, conversely, that the inclusion implies the two equalities for *any values* of u and v; and, in fact, we restrict ourselves to the proof that this implication holds for some value of u and v, namely for the particular values

$$u = a, \quad b = v;$$

for we have

$$(a = ab) = (a < b) = (a + b = b).$$

But we cannot conclude, from the fact that the implication (and therefore also the equivalence) is true for *some* value of the indeterminates, that it is true for *all*; in particular, it is not true for the values

$$u = 1, \quad v = 0,$$

for then $(a = bu)$ and $(a + v = b)$ become $(a = b)$, which obviously asserts more than the given inclusion $(a < b)$.[39]

Therefore we can write only the equivalences

$$(a < b) = \sum_u (a = bu) = \sum_v (a + v = b),$$

[39] Likewise if we make

$$u = 0, \quad v = 1,$$

we obtain the equalities

$$(a = 0), \quad (b = 1),$$

which assert still more than the given inclusion.

but the three expressions

$$(a < b), \quad \prod_u (a = bu), \quad \prod_v (a + v = b)$$

are not equivalent.[40]

0.35 The Expression of a Double Inclusion by Means of an Indeterminate

THEOREM. *The double inclusion*

$$b < x < a$$

is equivalent to the equality $x = au + bu'$ together with the condition $(b < a)$, u being a term absolutely indeterminate.

Demonstration.—Let us develop an equality in question,

$$x(a'u + b'u') + x'(au + bu') = 0,$$
$$(a'x + ax')u + (b'x + bx')u' = 0.$$

Eliminating u from it,

$$a'b'x + abx' = 0.$$

This equality is equivalent to the double inclusion

$$ab < x < a + b.$$

---

[40]According to the remark in the preceding note, it is clear that we have

$$\prod_v (a = bu) = (a = b = 0), \quad \prod_v (a + v = b) = (a = b = 1),$$

since the equalities affected by the sign $\prod$ may be likewise verified by the values

$$u = 0, \quad u = 1 \quad \text{and} \quad v = 0, \quad v = 1.$$

If we wish to know within what limits the indeterminates u and v are variable, it is sufficient to solve with respect to them the equations

$$(a < b) = (a = bu), \quad (a < b) = (a + v = b),$$

or

$$(ab' = a'bu + ab' + au', \quad ab' = ab' + b'v + a'bv',$$

or

$$a'bu + abu' = 0, \quad a'b'v + a'bv' = 0,$$

from which (by a formula to be demonstrated later on) we derive the solutions

$$u = ab + w(a + b'), \quad v = a'b + w(a + b),$$

or simply

$$u = ab + wb', \quad v = a'b + wa,$$

w being absolutely indeterminate. We would arrive at these solutions simply by asking: By what term must we multiply b in order to obtain a? By a term which contains ab plus any part of b'. What term must we add to a in order to obtain b? A term which contains $a'b$ plus any part of a. In short, u can vary between ab and $a + b'$, v between $a'b$ and $a + b$.

But, by hypothesis, we have

$$(b < a) = (ab = b) = (a + b = a).$$

The double inclusion is therefore reduced to

$$b < x < a.$$

So, whatever the value of u, the equality under consideration involves the double inclusion. Conversely, the double inclusion involves the equality, whatever the value of x may be, for it is equivalent to

$$a'x + bx' = 0,$$

and then the equality is simplified and reduced to

$$ax'u + b'xu' = 0.$$

We can always derive from this the value of u in terms of x, for the resultant $(ab'xx' = 0)$ is identically verified. The solution is given by the double inclusion

$$b'x < u < a' + x.$$

Remark.—There is no contradiction between this result, which shows that the value of u lies between certain limits, and the previous assertion that u is absolutely indeterminate; for the latter assumes that x is any value that will verify the double inclusion, while when we evaluate u in terms of x the value of x is supposed to be determinate, and it is with respect to this particular value of x that the value of u is subjected to limits.[41]

In order that the value of u should be completely determined, it is necessary and sufficient that we should have

$$b'x = a' + x,$$

that is to say,

$$b'xax' + (b + x')(a' + x) = 0$$

or

$$bx + a'x' = 0.$$

Now, by hypothesis, we already have

$$a'x + bx' = 0.$$

If we combine these two equalities, we find

$$(a + b = 0) = (a = 1)(b = 0).$$

[41]Moreover, if we substitute for x its inferior limit b in the inferior limit of u, this limit becomes $bb' = 0$; and, if we substitute for x its superior limit a in the superior limit of u, this limit becomes $a + a' = 1$.

This is the case when the value of x is absolutely indeterminate, since it lies between the limits 0 and 1.

In this case we have

$$u = b'x = a + x = x.$$

In order that the value of u be absolutely indeterminate, it is necessary and sufficient that we have at the same time

$$b'x = 0, \quad a' + x = 1,$$

or

$$b'x + ax' = 0,$$

that is

$$a < x < b.$$

Now we already have, by hypothesis,

$$b < x < a;$$

so we may infer

$$b = x = a.$$

This is the case in which the value of x is completely determinate.

0.36 Solution of an Equation Involving One Unknown Quantity

The solution of the equation

$$ax + bx' = 0$$

may be expressed in the form

$$x = a'u + bu',$$

u being an indeterminate, on condition that the resultant of the equation be verified; for we can prove that this equality implies the equality

$$ab'x + a'bx' = 0,$$

which is equivalent to the double inclusion

$$a'b < x < a' + b.$$

Now, by hypothesis, we have

$$(ab = 0) = (a'b = b) = (a' + b = a').$$

Therefore, in this hypothesis, the proposed solution implies the double inclusion

$$b < x < a';$$

which is equivalent to the given equation.

Remark.—In the same hypothesis in which we have

$$(ab = 0) = (b < a'),$$

we can always put this solution in the simpler but less symmetrical forms

$$x = b + a'u, \quad x = a'(b + u).$$

For

1. We have identically

$$b = bu + bu'.$$

Now

$$(b < a') < (bu < a'u).$$

Therefore

$$(x = bu' + a'u) = (x = b + a'u).$$

2. Let us now demonstrate the formula

$$x = a'b + a'u.$$

Now

$$a'b = b.$$

Therefore

$$x = b + a'u$$

which may be reduced to the preceding form.

Again, we can put the same solution in the form

$$x = a'b + u(ab + a'b'),$$

which follows from the equation put in the form

$$ab'x + a'bx' = 0,$$

if we note that

$$a' + b = ab + a'b + a'b'$$

and that

$$ua'b < a'b.$$

This last form is needlessly complicated, since, by hypothesis,

$$ab = 0.$$

Therefore there remains

$$x = a'b + ua'b'$$

which again is equivalent to

$$x = b + ua',$$

since

$$a'b = b \quad \text{and} \quad a' = a'b + a'b'.$$

Whatever form we give to the solution, the parameter u in it is absolutely indeterminate, *i.e.*, it can receive all possible values, including 0 and 1; for when $u = 0$ we have

$$x = b,$$

and when $u = 1$ we have

$$x = a',$$

and these are the two extreme values of x.

Now we understand that x is determinate in the particular case in which $a' = b$, and that, on the other hand, it is absolutely indeterminate when

$$b = 0, \quad a' = 1, \quad (\text{or } a = 0).$$

Summing up, the formula

$$x = a'u + bu'$$

replaces the "limited" variable x (lying between the limits a' and b) by the "unlimited" variable u which can receive all possible values, including 0 and 1.

Remark.[42]—The formula of solution

$$x = a'x + bx'$$

is indeed equivalent to the given equation, but not so the formula of solution

$$x = a'u + bu'$$

as a function of the indeterminate u. For if we develop the latter we find

$$ab'x + a'bx' + ab(xu + x'u') + a'b'(xu' + x'u) = 0,$$

and if we compare it with the developed equation

$$ab + ab'x + a'bx' = 0,$$

we ascertain that it contains, besides the solution, the equality

$$ab(xu' + x'u) = 0,$$

and lacks of the same solution the equality

$$a'b'(xu' + x'u) = 0.$$

Moreover these two terms disappear if we make

$$u = x$$

[42] Poretsky. *Sept lois*, Chaps. XXXIII and XXXIV.

and this reduces the formula to

$$x = a'x + bx'.$$

From this remark, PORETSKY concluded that, in general, the solution of an equation is neither a consequence nor a cause of the equation. It is a cause of it in the particular case in which

$$ab = 0,$$

and it is a consequence of it in the particular case in which

$$(a'b' = 0) = (a + b = i).$$

But if ab is not equal to 0, the equation is unsolvable and the formula of solution absurd, which fact explains the preceding paradox. If we have at the same time

$$ab = 0 \quad \text{and} \quad a + b = 1,$$

the solution is both consequence and cause at the same time, that is to say, it is equivalent to the equation. For when $a = b$ the equation is determinate and has only the one solution

$$x = a' = b.$$

Thus, whenever an equation is solvable, its solution is one of its causes; and, in fact, the problem consists in finding a value of x which will verify it, *i.e.*, which is a cause of it.

To sum up, we have the following equivalence:

$$(ax + bx' = 0) = (ab = 0) \sum_u (x = a'u + bu')$$

which includes the following implications:

$$(ax + bx' = 0) < (ab = 0),$$
$$(ax + bx' = 0) < \sum_u (x = a'u + bu'),$$
$$(ab = 0) \sum_u (x = a'u + bu') < (ax + bx' = 0).$$

0.37 Elimination of Several Unknown Quantities

We shall now consider an equation involving several unknown quantities and suppose it reduced to the normal form, *i.e.*, its first member developed with respect to the unknown quantities, and its second member zero. Let us first concern ourselves with the problem of elimination. We can eliminate the unknown quantities either one by one or all at once.

For instance, let

$$\phi(x, y, z) = axyz + bxyz' + cxy'z + dxy'z' + fx'yz + gx'yz' + hx'y'z + kx'y'z' = 0 \tag{5}$$

be an equation involving three unknown quantities.

We can eliminate z by considering it as the only unknown quantity, and we obtain as resultant

$$(axy + cxy' + fx'y + hx'y')(bxy + dxy' + gx'y + kx'y') = 0$$

or

$$abxy + cdxy' + fgx'y + hkx'y' = 0. \tag{6}$$

If equation (5) is possible, equation (6) is possible as well; that is, it is verified by some values of x and y. Accordingly we can eliminate y from the equation by considering it as the only unknown quantity, and we obtain as resultant

$$(abx + fgx')(cdx + hkx') = 0$$

or

$$abcdx + fghkx' = 0. \tag{7}$$

If equation (5) is possible, equation (7) is also possible;. that is, it is verified by some values of x. Hence we can eliminate x from it and obtain as the final resultant,

$$abcd \cdot fghk = 0$$

which is a consequence of (5), independent of the unknown quantities. It is evident, by the principle of symmetry, that the same resultant would be obtained if we were to eliminate the unknown quantities in a different order. Moreover this result might have been foreseen, for since we have (§0.28)

$$abcdfghk < \phi(x, y, z),$$

$\phi(x, y, z)$ can vanish only if the product of its coefficients is zero:

$$[\phi(x, y, z) = 0] < (abcdfghk = 0).$$

Hence we can eliminate all the unknown quantities at once by equating to 0 the product of the coefficients of the function developed with respect to all these unknown quantities.

We can also eliminate some only of the unknown quantities at one time. To do this, it is sufficient to develop the first member with respect to these unknown quantities and to equate the product of the coefficients of this development to 0. This product will generally contain the other unknown quantities. Thus the resultant of the elimination of z alone, as we have seen, is

$$abxy + cdxy' + fgx'y + hkx'y' = 0$$

and the resultant of the elimination of y and z is

$$abcdx + fghkx' = 0.$$

These partial resultants can be obtained by means of the following practical rule: Form the constituents relating to the unknown quantities to be retained; give each of them, for a coefficient, the product of the coefficients of the constituents of the general development of which it is a factor, and equate the sum to 0.

0.38 Theorem Concerning the Values of a Function

All the values which can be assumed by a function of any number of variables $f(x, y, z \ldots)$ are given by the formula

$$abc \ldots k + u(a + b + c + \ldots + k),$$

in which u is absolutely indeterminate, and $a, b, c \ldots, k$ are the coefficients of the development of f.

Demonstration.—It is sufficient to prove that in the equality

$$f(x, y, z \ldots) = abc \ldots k + u(a + b + c + \ldots + k)$$

u can assume all possible values, that is to say, that this equality, considered as an equation in terms of u, is indeterminate.

In the first place, for the sake of greater homogeneity, we may put the second member in the form

$$u'abc \ldots k + u(a + b + c + \ldots + k),$$

for

$$abc \ldots k = uabc \ldots k + u'abc \ldots k,$$

and

$$uabc \ldots k < u(a + b + c + \ldots + k).$$

Reducing the second member to 0 (assuming there are only three variables x, y, z)

$$\begin{aligned}(axyz + bxyz' + cxy'z + \ldots + kx'y'z') \\ \times [ua'b'c' \ldots k' + u'(a' + b' + c' + \ldots + k')] \\ + (a'xyz + b'xyz' + c'xy'z + \ldots + k'x'y'z') \\ \times [u(a + b + c + \ldots + k) + u'abc \ldots k] = 0,\end{aligned}$$

or more simply

$$\begin{aligned}u(a + b + c + \ldots + k)(a'xyz + b'xyz + c'xy'z + \ldots + k'x'y'z') \\ + u'(a' + b' + c' + \ldots + k')(axyz + bxyz + \ldots + kx'yz) = 0.\end{aligned}$$

If we eliminate all the variables x, y, z, but not the indeterminate u, we get the resultant

$$u(a+b+c+\ldots+k)a'b'c'\ldots k' \\ +u'(a'+b'+c'+\ldots+k')abc\ldots k=0.$$

Now the two coefficients of u and u' are identically zero; it follows that u is absolutely indeterminate, which was to be proved.[43]

From this theorem follows the very important consequence that a function of any number of variables can be changed into a function of a single variable without diminishing or altering its "variability".

Corollary.—A function of any number of variables can become equal to either of its limits.

For, if this function is expressed in the equivalent form

$$abc\ldots k+u(a+b+c+\ldots+k),$$

it will be equal to its minimum $(abc\ldots k)$ when $u=0$, and to its maximum $(a+b+c+\ldots+k)$ when $u=1$.

Moreover we can verify this proposition on the primitive form of the function by giving suitable values to the variables.

Thus a function can assume all values comprised between its two limits, including the limits themselves. Consequently, it is absolutely indeterminate when

$$abc\ldots k=0 \quad \text{and} \quad a+b+c+\ldots+k=1$$

at the same time, or

$$abc\ldots k=0=a'b'c'\ldots k'.$$

0.39 Conditions of Impossibility and Indetermination

The preceding theorem enables us to find the conditions under which an equation of several unknown quantities is impossible or indeterminate. Let $f(x, y, z\ldots)$ be the first member supposed to be developed, and $a, b, c\ldots, k$ its coefficients. The necessary and sufficient condition for the equation to be possible is

$$abc\ldots k=0.$$

For, (1) if f vanishes for some value of the unknowns, its inferior limit $abc\ldots k$ must be zero; (2) if $abc\ldots k$ is zero, f may become equal to it, and therefore may vanish for certain values of the unknowns.

The necessary and sufficient condition for the equation to be indeterminate (identically verified) is

$$a+b+c\ldots+k=0.$$

[43] Whitehead, *Universal Algebra*, Vol. I, §33 (4).

For, (1) if $a+b+c+\ldots+k$ is zero, since it is the superior limit of f, this function will always and necessarily be zero; (2) if f is zero for all values of the unknowns, $a+b+c+\ldots+k$ will be zero, for it is one of the values of f.

Summing up, therefore, we have the two equivalences

$$\sum[f(x,y,z,\ldots)=0]=(abc\ldots k=0).$$
$$\prod[f(x,y,z,\ldots)=0]=(a+b+c\ldots+k=0).$$

The equality $abc\ldots k=0$ is, as we know, the resultant of the elimination of all the unknowns; it is the consequence that can be derived from the equation (assumed to be verified) independently of all the unknowns.

0.40 Solution of Equations Containing Several Unknown Quantities

On the other hand, let us see how we can solve an equation with respect to its various unknowns, and, to this end, we shall limit ourselves to the case of two unknowns

$$axy+bxy'+cx'y+dx'y'=0.$$

First solving with respect to x,

$$x=(a'y+b'y')x+(cy+dy')x'.$$

The resultant of the elimination of x is

$$acy+bdy'=0.$$

If the given equation is true, this resultant is true.

Now it is an equation involving y only; solving it,

$$y=(a'+c')y+bdy'.$$

Had we eliminated y first and then x, we would have obtained the solution

$$y=(a'x+c'x')y+(bx+dx')y'$$

and the equation in x

$$abx+cdx'=0,$$

whence the solution

$$x=(a'+b')x+cdx'.$$

We see that the solution of an equation involving two unknown quantities is not symmetrical with respect to these unknowns; according to the order in which they were eliminated, we have the solution

$$x=(a'y+b'y')x+(cy+dy')x',$$
$$y=(a'+c')y+bdy',$$

or the solution

$$x = (a' + b')x + cdx,$$
$$y = (a'x + c'x')y + (bx + dx')y'.$$

If we replace the terms x, y, in the second members by indeterminates u, v, one of the unknowns will depend on only one indeterminate, while the other will depend on two. We shall have a symmetrical solution by combining the two formulas,

$$x = (a' + b')u + cdu',$$
$$y = (a' + c')v + bdv',$$

but the two indeterminates u and v will no longer be independent of each other. For if we bring these solutions into the given equation, it becomes

$$abcd + ab'c'uv + a'bd'uv' + a'cd'u'v + b'c'du'v' = 0$$

or since, by hypothesis, the resultant $abcd = 0$ is verified,

$$ab'c'uv + a'bd'uv' + a'cdu'v + b'c'du'v' = 0.$$

This is an "equation of condition" which the indeterminates u and v must verify; it can always be verified, since its resultant is identically true,

$$ab'c' \cdot a'bd' \cdot a'cd' \cdot b'c'd = aa' \cdot bb' \cdot cc' \cdot dd' = 0,$$

but it is not verified by any pair of values attributed to u and v.

Some general symmetrical solutions, *i.e.*, symmetrical solutions in which the unknowns are expressed in terms of several independent indeterminates, can however be found. This problem has been treated by SCHRÖDER [44], by WHITEHEAD [45] and by JOHNSON. [46]

This investigation has only a purely technical interest; for, from the practical point of view, we either wish to eliminate one or more unknown quantities (or even all), or else we seek to solve the equation with respect to one particular unknown. In the first case, we develop the first member with respect to the unknowns to be eliminated and equate the product of its coefficients to 0. In the second case we develop with respect to the unknown that is to be extricated and apply the formula for the solution of the equation of one unknown quantity. If it is desired to have the solution in terms of some unknown quantities or in terms of the known only, the other unknowns (or all the unknowns) must first be eliminated before performing the solution.

[44] *Algebra der Logik*, Vol. I, §24.
[45] *Universal Algebra*, Vol. I, §§35–37.
[46] "Sur la théorie des égalités logiques", *Bibl. du Cong. intern. de Phil.*, Vol. III, p. 185 (Paris, 1901).

0.41 The Problem of Boole

According to BOOLE the most general problem of the algebra of logic is the following[47]:

Given any equation (which is assumed to be possible)

$$f(x, y, z, \dots) = 0,$$

and, on the other hand, the expression of a term t in terms of the variables contained in the preceding equation

$$t = \varphi(x, y, z, \dots)$$

to determine the expression of t in terms of the constants contained in f and in φ.

Suppose f and φ developed with respect to the variables $x, y, z \dots$ and let $p_1, p_2, p_3, \dots$ be their constituents:

$$f(x, y, z, \dots) = Ap_1 + Bp_2 + Cp_3 + \dots,$$
$$\phi(x, y, z, \dots) = ap_1 + bp_2 + cp_3 + \dots.$$

Then reduce the equation which expresses t so that its second member will be 0:

$$(t\phi' + t'\phi = 0) = [(a'p_1 + b'p_2 + c'p_3 + \dots)t + (ap_1 + bp_2 + cp_3 + \dots)t' = 0].$$

Combining the two equations into a single equation and developing it with respect to t:

$$[(A + a')p_1 + (B + b')p_2 + (C + c')p_3 + \dots]t + [(A + a)p_1 + (B + b)p_2 + (C + c)p_3 + \dots]t' = 0.$$

This is the equation which gives the desired expression of t. Eliminating t, we obtain the resultant

$$Ap_1 + Bp_2 + Cp_3 + \dots = 0,$$

as we might expect. If, on the other hand, we wish to eliminate $x, y, z, \dots$ (*i.e.*, the constituents $p_1, p_2, p_3 \dots$), we put the equation in the form

$$(A + a't + at')p_1 + (B + b't + bt')p_2 + (C + c't + ct')p_3 + \dots = 0,$$

and the resultant will be

$$(A + a't + at')(B + b't + bt')(C + c't + ct') \dots = 0,$$

[47] *Laws of Thought*, Chap. IX, §8.

an equation that contains only the unknown quantity t and the constants of the problem (the coefficients of f and of φ). From this may be derived the expression of t in terms of these constants. Developing the first member of this equation

$$(A + a')(B + b')(C + c') \dots \times t + (A + a)(B + b)(C + c) \dots \times t' = 0.$$

The solution is

$$t = (A + a)(B + b)(C + c) \dots + u(A'a + B'b + C'c + \dots).$$

The resultant is verified by hypothesis since it is

$$ABC \dots = 0,$$

which is the resultant of the given equation

$$f(x, y, z, \dots) = 0.$$

We can see how this equation contributes to restrict the variability of t. Since t was defined only by the function φ, it was determined by the double inclusion

$$abc \dots < t < a + b + c + \dots.$$

Now that we take into account the condition $f = 0$, t is determined by the double inclusion

$$(A + a)(B + b)(C + c) \dots < t < (A'a + B'b + C'c + \dots).$$ [48]

The inferior limit can only have increased and the superior limit diminished, for

$$abc \dots < (A + a)(B + b)(C + c) \dots$$

and

$$A'a + B'b + C'c \dots < a + b + c \dots.$$

The limits do not change if $A = B = C = \dots = 0$, that is, if the equation $f = 0$ is reduced to an identity, and this was evident *a priori*.

0.42 The Method of Poretsky

The method of BOOLE and SCHRÖDER which we have heretofore discussed is clearly inspired by the example of ordinary algebra, and it is summed up in two processes analogous to those of algebra, namely the solution of equations with reference to unknown quantities and elimination of the unknowns. Of these processes the second is much the more important from a logical point of view, and BOOLE was even on the point of considering deduction as essentially consisting in the *elimination of middle terms*. This notion, which is too restricted,

[48] WHITEHEAD, *Universal Algebra*, p. 63.

was suggested by the example of the syllogism, in which the conclusion results from the elimination of the middle term, and which for a long time was wrongly considered as the only type of mediate deduction.[49]

However this may be, BOOLE and SCHRÖDER have exaggerated the analogy between the algebra of logic and ordinary algebra. In logic, the distinction of known and unknown terms is artificial and almost useless. All the terms are—in principle at least—known, and it is simply a question, certain relations between them being given, of deducing new relations (unknown or not explicitly known) from these known relations. This is the purpose of PORETSKY'S method which we shall now expound. It may be summed up in three laws, the *law of forms*, the *law of consequences* and the *law of causes*.

0.43 The Law of Forms

This law answers the following problem: An equality being given, to find for any term (simple or complex) a determination equivalent to this equality. In other words, the question is to find all the *forms* equivalent to this equality, any term at all being given as its first member.

We know that any equality can be reduced to a form in which the second member is 0 or 1; *i.e.*, to one of the two equivalent forms

$$N = 0, \qquad N' = 1.$$

The function N is what PORETSKY calls the *logical zero* of the given equality; N' is its logical *whole*.[50]

Let U be any term; then the determination of U:

$$U = N'U + NU'$$

is equivalent to the proposed equality; for we know it is equivalent to the equality

$$(NU + NU' = 0) = (N = 0).$$

Let us recall the signification of the determination

$$U = N'U + NU'.$$

[49] In fact, the fundamental formula of elimination

$$(ax + bx' = 0) < (ab = 0)$$

is, as we have seen, only another form and a consequence of the principle of the syllogism

$$(b < x < a') < (b < a').$$

[50] They are called "logical" to distinguish them from the identical *zero* and *whole*, *i.e.*, to indicate that these two terms are not equal to 0 and 1 respectively except by virtue of the data of the problem.

It denotes that the term U is contained in N' and contains N. This is easily understood, since, by hypothesis, N is equal to 0 and N' to 1. Therefore we can formulate the *law of forms* in the following way:

To obtain all the forms equivalent to a given equality, it is sufficient to express that any term contains the logical zero of this equality and is contained in its logical whole.

The number of forms of a given equality is unlimited; for any term gives rise to a form, and to a form different from the others, since it has a different first member. But if we are limited to the universe of discourse determined by n simple terms, the number of forms becomes finite and determinate. For, in this limited universe, there are 2^n constituents. Now, all the terms in this universe that can be conceived and defined are sums of some of these constituents. Their number is, therefore, equal to the number of combinations that can be made with 2^n constituents, namely 2^{2^n} (including 0, the combination of 0 constituent, and 1, the combination of all the constituents). This will also be the number of different forms of any equality in the universe in question.

0.44 The Law of Consequences

We shall now pass to the law of consequences. Generalizing the conception of BOOLE, who made deduction consist in the elimination of middle terms, PORETSKY makes it consist in the elimination of known terms (*connaissances*). This conception is explained and justified as follows.

All problems in which the data are expressed by logical equalities or inclusions can be reduced to a single logical equality by means of the formula[51]

$$(A = 0)(B = 0)(C = 0) \ldots = (A + B + C \ldots = 0).$$

In this logical equality, which sums up all the data of the problem, we develop the first member with respect to all the simple terms which appear in it (and not with respect to the unknown quantities). Let n be the number of simple terms; then the number of the constituents of the development of 1 is 2^n. Let m ($\leq 2^n$) be the number of those constituents appearing in the first member of the equality. All possible consequences of this equality (in the universe of the n terms in question) may be obtained by forming all the additive combinations of these m constituents, and equating them to 0; and this is done in virtue of the formula

$$(A + B = 0) < (A = 0).$$

We see that we pass from the equality to any one of its consequences by suppressing some of the constituents in its first member, which correspond to as many elementary equalities (having 0 for second member), *i.e.*, as many as there are data in the problem. This is what is meant by "eliminating the known terms".

[51]We employ capitals to denote complex terms (logical functions) in contrast to simple terms denoted by small letters $(a, b, c, \ldots)$

The number of consequences that can be derived from an equality (in the universe of n terms with respect to which it is developed) is equal to the number of additive combinations that may be formed with its m constituents; *i.e.*, 2^m. This number includes the combination of 0 constituents, which gives rise to the identity $0 = 0$, and the combination of the m constituents, which reproduces the given equality.

Let us apply this method to the equation with one unknown quantity

$$ax + bx' = 0.$$

Developing it with respect to the *three* terms a, b, x:

$$\begin{aligned}(abx + ab'x + abx' + a'bx' = 0)\\ = [ab(x + x') + ab'x + a'bx' = 0]\\ = (ab = 0)(ab'x = 0)(a'bx' = 0).\end{aligned}$$

Thus we find, on the one hand, the resultant $ab = 0$, and, on the other hand, two equalities which may be transformed into the inclusions

$$x < a' + b, \qquad a'b < x.$$

But by the resultant which is equivalent to $b < a'$, we have

$$a' + b' = a', \qquad a'b = b.$$

This consequence may therefore be reduced to the double inclusion

$$x < a', \qquad b < x,$$

that is, to the known solution.

Let us apply the same method to the premises of the syllogism

$$(a < b)(b < c).$$

Reduce them to a single equality

$$(a < b) = (ab' = 0), \quad (b < c) = (bc' = 0), \quad (ab' + bc' = 0),$$

and seek all of its consequences.

Developing with respect to the three terms a, b, c:

$$abc' + ab'c + ab'c' + a'bc' = 0.$$

The consequences of this equality, which contains four constituents, are 16 (2^4) in number as follows:

1. $(abc' = 0) = (ab < c)$;
2. $(ab'c = 0) = (ac < b)$;
3. $(ab'c' = 0) = (a < b + c)$;
4. $(a'bc' = 0) = (b < a + c)$;
5. $(abc' + ab'c = 0) = (a < bc + b'c')$;
6. $(abc' + ab'c' = 0) = (ac' = 0) = (a < c)$.

This is the traditional conclusion of the syllogism.[52]

7. $$(abc' + a'bc' = 0) = (bc' = 0) = (b < c).$$

This is the second premise.

8. $$(ab'c + ab'c' = 0) = (ab' = 0) = (a < b).$$

This is the first premise.

9. $$(ab'c + a'bc' = 0) = (ac < b < a + c);$$

10. $$(ab'c' + a'bc' = 0) = (ab' + a'b < c);$$

11. $$(abc' + ab'c + ab'c' = 0) = (ab' + ac' = 0) = (a < bc);$$

12. $$(abc' + ab'c + a'bc' = 0) = (ab'c + bc' = 0) = (ac < b < c);$$

13. $$(abc' + ab'c' + a'bc' = 0) = (ac' + bc' = 0) = (a + b < c);$$

14. $$(ab'c + ab'c' + a'bc' = 0) = (ab' + a'bc' = 0) = (a < b < a + c).$$

The last two consequences (15 and 16) are those obtained by combining 0 constituent and by combining all; the first is the identity

15. $$0 = 0,$$

which confirms the paradoxical proposition that the true (identity) is implied by any proposition (is a consequence of it); the second is the given equality itself

16. $$ab' + bc' = 0,$$

which is, in fact, its own consequence by virtue of the principle of identity. These two consequences may be called the "extreme consequences" of the proposed equality. If we wish to exclude them, we must say that the number of the consequences properly so called of an equality of m constituents is $2^m - 2$.

0.45 The Law of Causes

The method of finding the consequences of a given equality suggests directly the method of finding its *causes*, namely, the propositions of which it is the consequence. Since we pass from the cause to the consequence by eliminating known terms, *i.e.*, by suppressing constituents, we will pass conversely from the consequence to the cause by adjoining known terms, *i.e.*, by adding constituents to the given equality. Now, the number of constituents that may be added to

[52] It will be observed that this is the only consequence (except the two extreme consequences [see the text below]) independent of b; therefore it is the resultant of the elimination of that middle term.

it, *i.e.*, that do not already appear in it, is $2^n - m$. We will obtain all the possible causes (in the universe of the n terms under consideration) by forming all the additive combinations of these constituents, and adding them to the first member of the equality in virtue of the general formula

$$(A + B = 0) < (A = 0),$$

which means that the equality $(A = 0)$ has as its cause the equality $(A+B = 0)$, in which B is any term. The number of causes thus obtained will be equal to the number of the aforesaid combinations, or $2^{2n} - m$.

This method may be applied to the investigation of the causes of the premises of the syllogism

$$(a < b)(b < c)$$

which, as we have seen, is equivalent to the developed equality

$$abc' + ab'c + ab'c' + a'bc' = 0.$$

This equality contains four of the eight (2^3) constituents of the universe of three terms, the four others being

$$abc, a'bc, a'b'c, a'b'c'.$$

The number of their combinations is 16 (2^4), this is also the number of the causes sought, which are:

(8) $$(abc + abc' + ab'c + ab'c' + a'bc' = 0) \\ = (a + bc' = 0) = (a = 0)(b < c);$$

(9) $$(abc' + ab'c + ab'c' + a'bc + a'bc' = 0) \\ = (abc' + ab' + a'b = 0) = (ab < c)(a = b);$$

(10) $$(abc' + ab'c + ab'c' + a'bc' + a'b'c = 0) \\ = (bc' + b'c + ab'c' = 0) = (b = c)(a < b + c);$$

(11) $$(abc' + ab'c + ab'c' + a'bc' + a'b'c' = 0) \\ = (c' + ab' = 0) = (c = 1)(a < b);$$

(12) $$(abc + abc' + ab'c + ab'c' + a'bc + a'bc' = 0) \\ = (a + b = 0) = (a = 0)(b = 0);$$

(13) $$(abc + abc' + ab'c + ab'c' + a'bc' + a'b'c = 0) \\ = (a + bc' + b'c = 0) = (a = 0)(b = c);$$

(14) $$(abc + abc' + ab'c + ab'c' + a'bc' + a'b'c' = 0) \\ = (a + c' = 0) = (a = 0)(c = 1)^{53};$$

(15) $$(abc' + ab'c + ab'c' + a'bc + a'bc' + a'b'c = 0) \\ = (ac' + a'c + ab'c + a'bc' = 0) \\ = (a = c)(ac < b < a + c) = (a = b = c);$$

(16) $$(abc' + ab'c + ab'c' + a'bc + a'bc' + a'b'c = 0) \\ = (c' + ab' + a'b = 0) = (c = 1)(a = b);$$

(17) $$(abc' + ab'c + ab'c' + a'bc' + a'b'c + a'b'c' = 0) \\ = (b' + c' = 0) = (b = c = 1).$$

Before going any further, it may be observed that when the sum of certain constituents is equal to 0, the sum of the rest is equal to 1. Consequently, instead of examining the sum of seven constituents obtained by ignoring one of the four missing constituents, we can examine the equalities obtained by equating each of these constituents to 1:

(18) $$(a'b'c' = 1) = (a + b + c = 0) = (a = b = c = 0);$$

(19) $$(a'b'c = 1) = (a + b + c' = 0) = (a = b = 0)(c = 1);$$

(20) $$(a'bc = 1) = (a + b' + c' = 0) = (a = 0)(b = c = 1);$$

(21) $$(abc = 1) = (a = b = c = 1).$$

Note that the last four causes are based on the inclusion

$$0 < 1.$$

The last two causes (22 and 23) are obtained either by adding *all* the missing constituents or by not adding any. In the first case, the sum of all the constituents being equal to 1, we find

(22) $$1 = 0,$$

that is, absurdity, and this confirms the paradoxical proposition that the false (the absurd) implies any proposition (is its cause). In the second case, we obtain simply the given equality, which thus appears as one of its own causes (by the principle of identity):

$$ab' + bc' = 0. \tag{23}$$

If we disregard these two extreme causes, the number of causes properly so called will be

$$2^{2^n - m} - 2.$$

0.46 Forms of Consequences and Causes

We can apply the law of forms to the consequences and causes of a given equality so as to obtain all the forms possible to each of them. Since any equality is equivalent to one of the two forms

$$N = 0, \qquad N' = 1,$$

each of its consequences has the form[54]

$$NX = 0, \qquad \text{or } N' + X' = 1,$$

and each of its causes has the form

$$N + X = 0, \qquad \text{or } N'X' = 1.$$

In fact, we have the following formal implications:

$$(N + X = 0) < (N = 0) < (NX = 0),$$
$$(N'X' = 1) < (N' = 1) = (N' + X' = 1).$$

Applying the law of forms, the formula of the consequences becomes

$$U = (N' + X')U + NXU',$$

and the formula of the causes

$$U = N'X'U + (N + X)U';$$

[54]In §0.44 we said that a consequence is obtained by taking a part of the constituents of the first member N, and not by multiplying it by a term X; but it is easily seen that this amounts to the same thing. For, suppose that X (like N) be developed with respect to the n terms of discourse. It will be composed of a certain number of constituents. To perform the multiplication of N by X, it is sufficient to multiply all their constituents each by each. Now, the product of two identical constituents is equal to each of them, and the product of two different constituents is 0. Hence the product of N by X becomes reduced to the sum of the constituents common to N and X, which is, of course, contained in N. So, to multiply N by an arbitrary term is tantamount to taking a part of its constituents (or all, or none).

or, more generally, since X and X' are indeterminate terms, and consequently are not necessarily the negatives of each other, the formula of the consequences will be

$$U = (N' + X)U + NYU',$$

and the formula of the causes

$$U = N'XU + (N + Y)U'.$$

The first denotes that U is contained in $(N' + X)$ and contains NY; which indeed results, *a fortiori*, from the hypothesis that U is contained in N' and contains N.

The second formula denotes that U is contained in $N'X$ and contains $N' + Y$ whence results, *a fortiori*, that U is contained in N' and contains N.

We can express this rule verbally if we agree to call every class contained in another a ***sub-class***, and every class that contains another a ***super-class***. We then say: To obtain all the consequences of an equality (put in the form $U = N'U + NU'$), it is sufficient to substitute for its logical whole N' all its super-classes, and, for its logical zero N, all its sub-classes. Conversely, to obtain all the causes of the same equality, it is sufficient to substitute for its logical whole all its sub-classes, and for its logical zero, all its super-classes.

0.47 Example: Venn's Problem

The members of the administrative council of a financial society are either bondholders or shareholders, but not both. Now, all the bondholders form a part of the council. What conclusion must we draw?

Let a be the class of the members of the council; let b be the class of the bondholders and c that of the shareholders. The data of the problem may be expressed as follows:

$$a < bc' + b'c, \qquad b < a.$$

Reducing to a single developed equality,

$$a(bc = b'c') = 0, \qquad a'b = 0,$$

(24) $$abc + ab'c' + a'bc + a'bc' = 0.$$

This equality, which contains 4 of the constituents, is equivalent to the following, which contains the four others,

(25) $$abc' + ab'c + a'b'c + a'b'c' = 1.$$

This equality may be expressed in as many different forms as there are classes in the universe of the three terms a, b, c.

Ex. 1. $$a = abc' + ab'c + a'bc + a'bc',$$

that is,

$$b < a < bc' + b'c,$$

Ex. 2. $$b = abc' + ab'c = ac';$$

Ex. 3. $$c = ab'c + a'b'c + ab'c' + a'bc'$$

that is,

$$ab' + a'b < c < b'.$$

These are the solutions obtained by solving equation (24) with respect to a, b, and c.

From equality (24) we can derive 16 consequences as follows :

1. $$abc = 0;$$
2. $$(ab'c' = 0) = (a < b + c);$$
3. $$(a'bc = 0) = (bc < a);$$
4. $$(a'bc = 0) = (b < a + c);$$
5. $$(abc + ab'c' = 0) = (a < bc' + b'c) \text{ [1}^{\text{st}} \text{ premise]};$$
6. $$(abc + a'bc = 0) = (bc = 0);$$
7. $$(abc + a'bc' = 0) = (b < ac' + a'c);$$
8. $$(ab'c' + a'bc = 0) = (bc < a < b + c);$$
9. $$(ab'c' + a'bc' = 0) = (ab' + a'b < c);$$
10. $$(a'bc + a'bc' = 0) = (a'b = 0) \text{ [2}^{\text{d}} \text{ premise]};$$
11. $$(abc + ab'c' + a'bc = 0) = (bc + ab'c' = 0);$$
12. $$abc + ab'c + a'bc' = 0;$$
13. $$(abc + a'bc + a'bc' = 0) = (bc + a'bc') = 0;$$
14. $$ab'c' + a'bc + a'bc' = 0.$$

The last two consequences, as we know, are the identity $(0 = 0)$ and the equality (24) itself. Among the preceding consequences will be especially noted the 6th $(bc = 0)$, the resultant of the elimination of a, and the 10th $(a'b = 0)$, the resultant of the elimination of c. When b is eliminated the resultant is the identity

$$[(a' + c)ac' = 0] = (0 = 0).$$

Finally, we can deduce from the equality (24) or its equivalent (25) the following 16 causes:

1. $$(abc' = 1) = (a = 1)(b = 1)(c = 0);$$
2. $$(ab'c = 1) = (a = 1)(b = 0)(c = 1);$$
3. $$(a'b'c = 1) = (a = 0)(b = 0)(c = 1);$$
4. $$(a'b'c' = 1) = (a = 0)(b = 0)(c = 0);$$
5. $$(abc' + ab'c = 1) = (a = 1)(b' = c);$$
6. $$(abc' + a'b'c = 1) = (a = b = c');$$
7. $$(abc' + a'b'c' = 1) = (c = 0)(a = b);$$
8. $$(ab'c + a'b'c = 1) = (b = 0)(c = 1);$$
9. $$(ab'c + a'b'c' = 1) = (b = 0)(a = c);$$
10. $$(a'b'c + a'b'c' = 1) = (a = 0)(b = 0);$$
11. $$(abc' + ab'c + a'b'c = 1) = (b = c')(c' < a);$$
12. $$(abc' + ab'c + a'b'c' = 1) = (bc = 0)(a = b + c);$$
13. $$(abc' + a'b'c + a'b'c' = 1) = (ac = 0)(a = b);$$
14. $$(ab'c + a'b'c + a'b'c' = 1) = (b = o)(a < c).$$

The last two causes, as we know, are the equality (24) itself and the absurdity $(1 = 0)$. It is evident that the cause independent of a is the 8th $(b = 0)(c = 1)$, and the cause independent of c is the 10th $(a = 0)(b = 0)$. There is no cause, properly speaking, independent of b. The most "natural" cause, the one which may be at once divined simply by the exercise of common sense, is the 12th:

$$(bc = 0)(a = b + c).$$

But other causes are just as possible; for instance the 9th $(b = 0)(a = c)$, the 7th $(c = 0)(a = b)$, or the 13th $(ac = 0)(a = b)$.

We see that this method furnishes the complete enumeration of all possible cases. In particular, it comprises, among the *forms* of an equality, the solutions deducible therefrom with respect to such and such an "unknown quantity", and, among the *consequences* of an equality, the resultants of the elimination of such and such a term.

0.48 The Geometrical Diagrams of Venn

PORETSKY'S method may be looked upon as the perfection of the methods of STANLEY JEVONS and VENN.

Conversely, it finds in them a geometrical and mechanical illustration, for VENN'S method is translated in geometrical diagrams which represent all the constituents, so that, in order to obtain the result, we need only strike out (by shading) those which are made to vanish by the data of the problem. For instance, the universe of three terms a, b, c, represented by the unbounded plane,

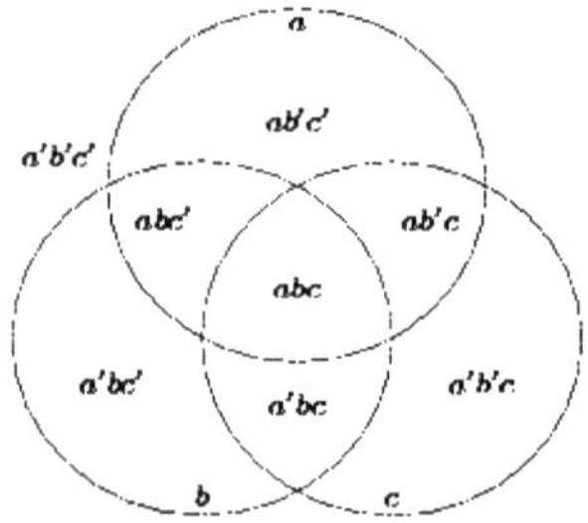

Figure 1:

is divided by three simple closed contours into eight regions which represent the eight constituents (Fig. 1).

To represent geometrically the data of VENN'S problem we must strike out the regions abc, $ab'c'$, $a'bc$ and $a'bc'$; there will then remain the regions abc', $ab'c$, $a'b'c$, and $a'b'c'$ which will constitute the universe *relative to the problem*, being what PORETSKY calls his *logical whole* (Fig. 2). Then every class will be contained in this universe, which will give for each class the expression resulting from the data of the problem. Thus, simply by inspecting the diagram, we see that the region bc does not exist (being struck out); that the region b is reduced to $ab'c'$ (hence to ab); that all a is b or c, and so on.

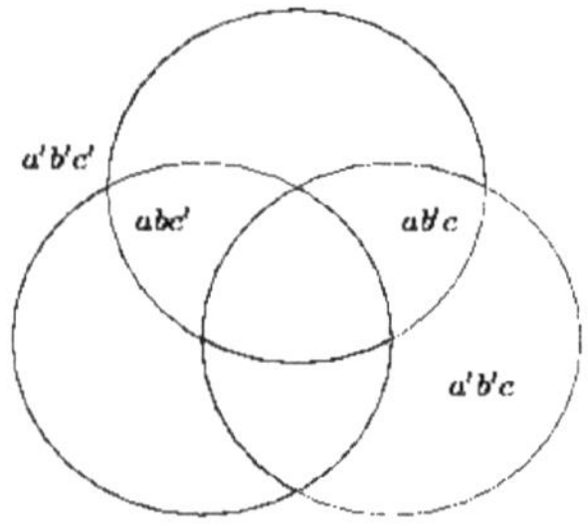

Figure 2:

This diagrammatic method has, however, serious inconveniences as a method for solving logical problems. It does not show how the data are exhibited by canceling certain constituents, nor does it show how to combine the remaining constituents so as to obtain the consequences sought. In short, it serves only to exhibit one single step in the argument, namely the equation of the problem; it dispenses neither with the previous steps, *i.e.*, "throwing of the problem into an equation" and the transformation of the premises, nor with the subsequent steps, *i.e.*, the combinations that lead to the various consequences. Hence it is of very little use, inasmuch as the constituents can be represented by algebraic symbols quite as well as by plane regions, and are much easier to deal with in this form.

0.49 The Logical Machine of Jevons

In order to make his diagrams more tractable, VENN proposed a mechanical device by which the plane regions to be struck out could be lowered and caused to disappear. But JEVONS invented a more complete mechanism, a sort of *logical piano*. The keyboard of this instrument was composed of keys indicating the various simple terms (a, b, c, d), their negatives, and the signs $+$ and $=$. Another part of the instrument consisted of a panel with movable tablets on which were written all the combinations of simple terms and their negatives; that is, all the constituents of the universe of discourse. Instead of writing out the equalities which represent the premises, they are "played" on a keyboard like that of a typewriter. The result is that the constituents which vanish because of the premises disappear from the panel. When all the premises have been "played", the panel shows only those constituents whose sum is equal to 1, that is, forms the universe with respect to the problem, its logical whole. This mechanical method has the advantage over VENN'S geometrical method of performing automatically the "throwing into an equation", although the premises must first be expressed in the form of equalities; but it throws no more light than the geometrical method on the operations to be performed in order to draw the consequences from the data displayed on the panel.

0.50 Table of Consequences

But PORETSKY'S method can be illustrated, better than by geometrical and mechanical devices, by the construction of a table which will exhibit directly all the consequences and all the causes of a given equality. (This table is relative to this equality and each equality requires a different table). Each table comprises the 2^n classes that can be defined and distinguished in the universe of discourse of n terms. We know that an equality consists in the annulment of a certain number of these classes, viz., of those which have for constituents some of the constituents of its *logical zero* N. Let m be the number of these latter constituents, then the number of the subclasses of N is 2^m which, therefore, is the number of classes of the universe which vanish in consequence of the equality considered. Arrange them in a column commencing with 0 and ending with N (the two extremes). On the other hand, given any class at all, any preceding class may be added to it without altering its value, since by hypothesis they are null (in the problem under consideration). Consequently, by the data of the problem, each class is equal to 2^m classes (including itself). Thus, the assemblage of the 2^n classes of discourse is divided into 2^{n-m} series of 2^m classes, each series being constituted by the sums of a certain class and of the 2^m classes of the first column (sub-classes of N). Hence we can arrange these 2^m sums in the following columns by making them correspond horizontally to the classes of the first column which gave rise to them. Let us take, for instance, the very simple equality $a = b$, which is equivalent to

$$ab' + a'b = 0.$$

The logical zero (N) in this case is $ab' + a'b$. It comprises two constituents and consequently four sub-classes: 0, ab', $a'b$, and $ab' + a'b$. These will compose the first column. The other classes of discourse are ab, $a'b'$, $ab + a'b'$, and those obtained by adding to each of them the four classes of the first column. In this way, the following table is obtained:

$$\begin{array}{cccc}
0 & ab & a'b' & ab + a'b' \\
ab' & a & b' & a + b' \\
a'b & b & a' & a' + b \\
ab' + a'b & a + b & a' + b' & 1
\end{array}$$

By construction, each class of this table is the sum of those at the head of its row and of its column, and, by the data of the problem, it is equal to each of those in the same column. Thus we have 64 different consequences for any equality in the universe of discourse of 2 letters. They comprise 16 identities (obtained by equating each class to itself) and 16 forms of the given equality, obtained by equating the classes which correspond in each row to the classes which are known to be equal to them, namely

$$\begin{array}{cccc}
0 = ab' + a'b, & ab = a + b, & a'b' = a' + b' & ab + a'b' = 1 \\
a = b, & b' = a', & ab' = a'b, & a + b' = a' + b.
\end{array}$$

Each of these 8 equalities counts for two, according as it is considered as a determination of one or the other of its members.

0.51 Table of Causes

The same table may serve to represent all the causes of the same equality in accordance with the following theorem:

When the consequences of an equality $N = 0$ are expressed in the form of determinations of any class U, the causes of this equality are deduced from the consequences of the *opposite* equality, $N = 1$, put in the same form, by changing U to U' in one of the two members.

For we know that the consequences of the equality $N = 0$ have the form

$$U = (N' + X)U + NYU',$$

and that the causes of the same equality have the form

$$U = N'XU + (N + Y)U'.$$

Now, if we change U into U' in one of the members of this last formula, it becomes

$$U = (N + X')U + N'Y'U',$$

and the accents of X and Y can be suppressed since these letters represent indeterminate classes. But then we have the formula of the consequences of the equality $N' = 0$ or $N = 1$.

This theorem being established, let us construct, for instance, the table of causes of the equality $a = b$. This will be the table of the consequences of the opposite equality $a = b'$, for the first is equivalent to

$$ab' + a'b = 0,$$

and the second to

$$(ab + a'b' = 0) = (ab' + a'b = 1).$$

$$\begin{array}{cccc} 0 & ab' & a'b & ab' + a'b \\ ab & a & b & a + b \\ a'b' & b' & a' & a' + b' \\ ab + a'b' & a + b' & a' + b & 1 \end{array}$$

To derive the causes of the equality $a = b$ from this table instead of the consequences of the opposite equality $a = b'$, it is sufficient to equate the negative of each class to each of the classes in the same column. Examples are:

$$\begin{array}{lll} a' + b' = 0, & a' + b' = a'b', & a' + b' = ab + a'b', \\ a' + b = a, & a' + b = b', & a' + b = a + b'; \ldots. \end{array}$$

Among the 64 causes of the equality under consideration there are 16 absurdities (consisting in equating each class of the table to its negative); and 16 forms of the equality (the same, of course, as in the table of consequences, for two equivalent equalities are at the same time both cause and consequence of each other).

It will be noted that the table of causes differs from the table of consequences only in the fact that it is symmetrical to the other table with respect to the principal diagonal $(0, 1)$; hence they can be made identical by substituting the word "row" for the word "column" in the foregoing statement. And, indeed, since the rule of the consequences concerns only classes of the same column, we are at liberty so to arrange the classes in each column on the rows that the rule of the causes will be verified by the classes in the same row.

It will be noted, moreover, that, by the method of construction adopted for this table, the classes which are the negatives of each other occupy positions symmetrical with respect to the center of the table. For this result, the subclasses of the class N' (the logical whole of the given equality or the logical zero of the opposite equality) must be placed in the first row in their natural order from 0 to N'; then, in each division, must be placed the sum of the classes at the head of its row and column.

With this precaution, we may sum up the two rules in the following practical statement:

To obtain every consequence of the given equality (to which the table relates) it is sufficient to equate each class to every class in the same column; and, to obtain every cause, it is sufficient to equate each class to every class in the row occupied by its symmetrical class.

It is clear that the table relating to the equality $N = 0$ can also serve for the opposite equality $N = 1$, on condition that the words "row" and "column" in the foregoing statement be interchanged.

Of course the construction of the table relating to a given equality is useful and profitable only when we wish to enumerate all the consequences or the causes of this equality. If we desire only one particular consequence or cause relating to this or that class of the discourse, we make use of one of the formulas given above.

0.52 The Number of Possible Assertions

If we regard logical functions and equations as developed with respect to *all* the letters, we can calculate the number of assertions or different problems that may be formulated about n simple terms. For all the functions thus developed can contain only those constituents which have the coefficient 1 or the coefficient 0 (and in the latter case, they do not contain them). Hence they are additive combinations of these constituents; and, since the number of the constituents is 2^n, the number of possible functions is 2^{2^n}. From this must be deducted the function in which all constituents are absent, which is identically 0, leaving $2^{2^n} - 1$ possible equations (255 when $n = 3$). But these equations, in their turn, may be combined by logical addition, *i.e.*, by alternation; hence the number of their combinations is $2^{2^{2^n}-1} - 1$, excepting always the null combination. This is the number of possible assertions affecting n terms. When $n = 2$, this number is as high as 32767.[55] We must observe that only universal premises are admitted in this calculus, as will be explained in the following section.

0.53 Particular Propositions

Hitherto we have only considered propositions with an *affirmative* copula (*i.e.*, inclusions or equalities) corresponding to the *universal* propositions of classi-

[55] G. PEANO, *Calcolo geometrico* (1888) p. x; SCHRÖDER, *Algebra der Logik*, Vol. II, p. 144–148.

cal logic.[56] It remains for us to study propositions with a *negative* copula (non inclusions or inequalities), which translate *particular* propositions[57]; but the calculus of propositions having a negative copula results from laws already known, especially from the formulas of De Morgan and the law of contraposition. We shall enumerate the chief formulas derived from it.

The principle of composition gives rise to the following formulas:

$$(c \not< ab) = (c \not< a) + (c \not< b),$$
$$(a + b \not< c) = (a \not< c) + (b \not< c),$$

whence come the particular instances

$$(ab \neq 1) = (a \neq 1) + (b \neq 1),$$
$$(a + b \neq 0) = (a \neq 0) + (c \neq 0).$$

From these may be deduced the following important implications:

$$(a \neq 0) < (a + b \neq 0),$$
$$(a \neq 1) < (ab \neq 1).$$

From the principle of the syllogism, we deduce, by the law of transposition,

$$(a < b)(a \neq 0) < (b \neq 0),$$
$$(a < b)(b \neq 1) < (a \neq 1).$$

The formulas for transforming inclusions and equalities give corresponding formulas for the transformation of non-inclusions and inequalities,

$$(a \not< b) = (ab' \neq 0) = (a' + b \neq 1),$$
$$(a \neq b) = (ab' + a'b \neq 0) = (ab + a'b' + 1).$$

---

[56]The *universal affirmative*, "All a's are b's", may be expressed by the formulas

$$(a < b) = (a = ab) = (ab' = 0) = (a' + b = 1),$$

and the *universal negative*, "No a's are b's", by the formulas

$$(a < b') = (a = ab') = (ab = 0) = (a' + b' = 1).$$

[57]For the *particular affirmative*, "Some a's are b's", being the negation of the universal negative, is expressed by the formulas

$$(a \not< b') = (a \neq ab') = (ab \neq 0) = (a' + b' \neq 1),$$

and the *particular negative*, "Some a's are not b's", being the negation of the universal affirmative, is expressed by the formulas

$$(a \not< b) = (a \neq ab) = (ab' \neq 0) = (a' + b \neq 1).$$

0.54 Solution of an Inequation with One Unknown

If we consider the conditional inequality (*inequation*) with one unknown

$$ax + bx \neq 0,$$

we know that its first member is contained in the sum of its coefficients

$$ax + bx' < a + b.$$

From this we conclude that, if this inequation is verified, we have the inequality

$$a + b \neq 0.$$

This is the necessary condition of the solvability of the inequation, and the resultant of the elimination of the unknown x. For, since we have the equivalence

$$\prod_x (ax + bx' = 0) = (a + b = 0),$$

we have also by contraposition the equivalence

$$\sum_x (ax + bx' \neq 0) = (a + b \neq 0).$$

Likewise, from the equivalence

$$\sum_x (ax + bx' = 0) = (ab = 0)$$

we can deduce the equivalence

$$\prod_x (ax + bx' \neq 0) = (ab \neq 0),$$

which signifies that the necessary and sufficient condition for the inequation to be always true is

$$(ab \neq 0);$$

and, indeed, we know that in this case the equation

$$(ax + bx' = 0)$$

is impossible (never true).

Since, moreover, we have the equivalence

$$(ax + bx' = 0) = (x = a'x + bx'),$$

we have also the equivalence

$$(ax + bx' \neq 0) = (x \neq a'x + bx').$$

Notice the significance of this solution:

$$(ax + bx' \neq 0) = (ax \neq 0) + (bx' \neq 0) = (x \not< a') + (b \not< x).$$

"Either x is not contained in a', or it does not contain b". This is the negative of the double inclusion

$$b < x < a.$$

Just as the product of several equalities is reduced to one single equality, the sum (the alternative) of several inequalities may be reduced to a single inequality. But neither several alternative equalities nor several simultaneous inequalities can be reduced to one.

0.55 System of an Equation and an Inequation

We shall limit our study to the case of a simultaneous equality and inequality. For instance, let the two premises be

$$(ax + bx' = 0) \quad (cx + dx' \neq 0).$$

To satisfy the former (the equation) its resultant $ab = 0$ must be verified. The solution of this equation is

$$x = a'x + bx'.$$

Substituting this expression (which is equivalent to the equation) in the inequation, the latter becomes

$$(a'c + ad)x + (bc + b'd)x' \neq 0.$$

Its resultant (the condition of its solvability) is

$$(a'c + ad + bc + b'd \neq 0) = [(a' + b)c + (a + b')d \neq 0],$$

which, taking into account the resultant of the equality,

$$(ab = 0) = (a' + b = a') = (a + b' = b')$$

may be reduced to

$$a'c + b'd \neq 0.$$

The same result may be reached by observing that the equality is equivalent to the two inclusions

$$(x < a')(x' < b'),$$

and by multiplying both members of each by the same term

$$(cx < a'c)(dx' < b'd) < (cx + dx' < a'c + b'd)$$
$$(cx + dx' \neq 0) < (a'c + b'd \neq 0).$$

This resultant implies the resultant of the inequality taken alone

$$c + d \neq 0,$$

so that we do not need to take the latter into account. It is therefore sufficient to add to it the resultant of the equality to have the complete resultant of the proposed system

$$(ab = 0)(a'c + b'd \neq 0).$$

The solution of the transformed inequality (which consequently involves the solution of the equality) is

$$x \neq (a'c' + ad')x + (bc + b'd)x'.$$

0.56 Formulas Peculiar to the Calculus of Propositions.

All the formulas which we have hitherto noted are valid alike for propositions and for concepts. We shall now establish a series of formulas which are valid only for propositions, because all of them are derived from an axiom peculiar to the calculus of propositions, which may be called the *principle of assertion.*

This axiom is as follows:

Ax. 10

$$(a = 1) = a.$$

P. I.: To say that a proposition a is true is to state the proposition itself. In other words, to state a proposition is to affirm the truth of that proposition.[58]

Corollary:

$$a' = (a' = 1) = (a = 0).$$

P. I.: The negative of a proposition a is equivalent to the affirmation that this proposition is false.

By Ax. 9 (§0.20), we already have

$$(a = 1)(a = 0) = 0,$$

"A proposition cannot be both true and false at the same time", for

(Syll.) $$(a = 1)(a = 0) < (1 = 0) = 0.$$

But now, according to Ax. 10, we have

$$(a = 1) + (a = 0) = a + a' = 1.$$

[58]We can see at once that this formula is not susceptible of a conceptual interpretation (C. I.); for, if a is a concept, $(a = 1)$ is a proposition, and we would then have a logical equality (identity) between a concept and a proposition, which is absurd.

"A proposition is either true or false". From these two formulas combined we deduce directly that the propositions $(a = 1)$ and $(a = 0)$ are contradictory, *i.e.*,

$$(a \neq 1) = (a = 0), \qquad (a \neq 0) = (a = 1).$$

From the point of view of calculation Ax. 10 makes it possible to reduce to its first member every equality whose second member is 1, and to transform inequalities into equalities. Of course these equalities and inequalities must have propositions as their members. Nevertheless all the formulas of this section are also valid for classes in the particular case where the universe of discourse contains only one element, for then there are no classes but 0 and 1. In short, the special calculus of propositions is equivalent to the calculus of classes when the classes can possess only the two values 0 and 1.

0.57 Equivalence of an Implication and an Alternative

The fundamental equivalence

$$(a < b) = (a' + b = 1)$$

gives rise, by Ax. 10, to the equivalence

$$(a < b) = (a' + b)$$

which is no less fundamental in the calculus of propositions. To say that a implies b is the same as affirming "not-a or b", *i.e.*, "either a is false or b is true." This equivalence is often employed in every day conversation.

Corollary.—For any equality, we have the equivalence

$$(a = b) = ab + a'b'.$$

Demonstration:

$$(a = b) = (a < b)(b < a) = (a' + b)(b' + a) = ab + a'b'$$

"To affirm that two propositions are equal (equivalent) is the same as stating that either both are true or both are false".

The fundamental equivalence established above has important consequences which we shall enumerate.

In the first place, it makes it possible to reduce secondary, tertiary, etc., propositions to primary propositions, or even to sums (alternatives) of elementary propositions. For it makes it possible to suppress the copula of any proposition, and consequently to lower its order of complexity. An implication $(A < B)$, in which A and B represent propositions more or less complex, is reduced to the sum $A' + B$, in which only copulas within A and B appear, that is, propositions of an inferior order. Likewise an equality $(A = B)$ is reduced to the sum $(AB + A'B')$ which is of a lower order.

We know that the principle of composition makes it possible to combine several *simultaneous* inclusions or equalities, but we cannot combine alternative inclusions or equalities, or at least the result is not equivalent to their alternative but is only a consequence of it. In short, we have only the *implications*

$$(a < c) + (b < c) < (ab < c),$$
$$(c < a) + (c < b) < (c < a + b),$$

which, in the special cases where $c = 0$ and $c = 1$, become

$$(a = 0) + (b = 0) < (ab = 0),$$
$$(a = 1) + (b = 1) < (a + b = 1).$$

In the calculus of classes, the converse implications are not valid, for, from the statement that the class ab is null, we cannot conclude that one of the classes a or b is null (they can be not-null and still not have any element in common); and from the statement that the sum $(a + b)$ is equal to 1 we cannot conclude that either a or b is equal to 1 (these classes can *together* comprise all the elements of the universe without any of them *alone* comprising all). But these converse implications are true in the calculus of propositions

$$(ab < c) < (a < c) + (b < c),$$
$$(c < a + b) < (c < a) + (c < b);$$

for they are deduced from the equivalence established above, or rather we may deduce from it the corresponding equalities which imply them,

(1) $$(ab < c) = (a < c) + (b < c),$$
(2) $$(c < a + b) = (c < a) + (c < b).$$

Demonstration:

(1) $$(ab < c) = a' + b' + c,$$
$$(a < c) + (b < c) = (a' + c) + (b' + c) = a' + b' + c;$$
(2) $$(c < a + b) = c' + a + b,$$
$$(c < a) + (c < b) = (c' + a) + (c' + b) = c' + a + b.$$

In the special cases where $c = 0$ and $c = 1$ respectively, we find

(3) $$(ab = 0) = (a = 0) + (b = 0),$$
(4) $$(a + b = 1) = (a = 1) + (b = 1).$$

P. I.: (1) To say that two propositions united imply a third is to say that one of them implies this third proposition.

(2) To say that a proposition implies the alternative of two others is to say that it implies one of them.

(3) To say that two propositions combined are false is to say that one of them is false.

(4) To say that the alternative of two propositions is true is to say that one of them is true.

The paradoxical character of the first three of these statements will be noted in contrast to the self-evident character of the fourth. These paradoxes are explained, on the one hand, by the special axiom which states that a proposition is either true or false; and, on the other hand, by the fact that the false implies the true and that *only* the false is not implied by the true. For instance, if both premises in the first statement are true, each of them implies the consequence, and if one of them is false, it implies the consequence (true or false). In the second, if the alternative is true, one of its terms must be true, and consequently will, like the alternative, be implied by the premise (true or false). Finally, in the third, the product of two propositions cannot be false unless one of them is false, for, if both were true, their product would be true (equal to 1).

0.58 Law of Importation and Exportation

The fundamental equivalence $(a < b) = a' + b$ has many other interesting consequences. One of the most important of these is *the law of importation and exportation*, which is expressed by the following formula:

$$[a < (b < c)] = (ab < c).$$

"To say that if a is true b implies c, is to say that a and b imply c".

This equality involves two converse implications: If we infer the second member from the first, we *import* into the implication $(b < c)$ the hypothesis or condition a; if we infer the first member from the second, we, on the contrary, *export* from the implication $(ab < c)$ the hypothesis a.

Demonstration:

$$[a < (b < c)] = a' + (b < c) = a' + b' + c,$$
$$(ab < c) = (ab)' + c = a' + b' + c.$$

Cor. 1.—Obviously we have the equivalence

$$[a < (b < c)] = [b < (a < c)],$$

since both members are equal to $(ab < c)$, by the commutative law of multiplication.

Cor. 2.—We have also

$$[a < (a < b)] = (a < b),$$

for, by the law of importation and exportation,

$$[a < (a < b)] = (aa < b) = (a < b).$$

If we apply the law of importation to the two following formulas, of which the first results from the principle of identity and the second expresses the principle of contraposition,

$$(a < b) < (a < b), \qquad (a < b) < (b' < a'),$$

we obtain the two formulas

$$(a < b)a < b), \qquad (a < b)b' < a',$$

which are the two types of *hypothetical reasoning*: "If a implies b, and if a is true, b is true" (*modus ponens*); "If a implies b, and if b is false, a is false" (*modus tollens*).

Remark. These two formulas could be directly deduced by the principle of assertion, from the following

$$(a < b)(a = 1) < (b = 1),$$
$$(a < b)(b = 0) < (a = 0),$$

which are not dependent on the law of importation and which result from the principle of the syllogism.

From the same fundamental equivalence, we can deduce several paradoxical formulas:

1. $$a < (b < a), \qquad a' < (a < b).$$

"If a is true, a is implied by any proposition b; if a is false, a implies any proposition b". This agrees with the known properties of 0 and 1.

2. $$a < [(a < b) < b], \qquad a' < [(b < a) < b'].$$

"If a is true, then 'a implies b' implies b; if a is false, then 'b implies a' implies not-b."

These two formulas are other forms of hypothetical reasoning (*modus ponens* and *modus tollens*).

3. $$[(a < b) < a] = a,\text{[59]} \qquad [(b < a) < a'] = a',$$

"To say that, if a implies b, a is true, is the same as affirming a; to say that, if b implies a, a is false, is the same as denying a".

Demonstration:

$$[(a < b) < a] = (a' + b < a) = ab' + a = a,$$
$$[(b < a) < a'] = (b' + a < a') = a'b + a' = a'.$$

[59]This formula is BERTRAND RUSSELL'S "principle of reduction". See *The Principles of Mathematics*, Vol. I, p. 17 (Cambridge, 1903).

In formulas (1) and (3), in which b is any term at all, we might introduce the sign $\prod$ with respect to b. In the following formula, it becomes necessary to make use of this sign.

4. $$\prod_x \{[a < (b < x)] < x\} = ab.$$

Demonstration:

$$\begin{aligned}\{[a < (b < x)] < x\} &= \{[a' + (b < x)] < x\} \\ &= [(a' + b' + x) < x] = abx' + x = ab + x.\end{aligned}$$

We must now form the product $\prod_x (ab+x)$, where x can assume every value, including 0 and 1. Now, it is clear that the part common to all the terms of the form $(ab + x)$ can only be ab. For, (1) ab is contained in each of the sums $(ab+x)$ and therefore in the part common to all; (2) the part common to all the sums $(ab + x)$ must be contained in $(ab + 0)$, that is, in ab. Hence this common part is equal to ab,[60] which proved the theorem.

0.59 Reduction of Inequalities to Equalities

As we have said, the principle of assertion enables us to reduce inequalities to equalities by means of the following formulas:

$$(a \neq 0) = (a = 1), \qquad (a \neq 1) = (a = 0),$$
$$(a \neq b) = (a = b').$$

For,
$$(a \neq b) = (ab' + a'b + 0) = (ab' + ab' = 1) = (a = b').$$

Consequently, we have the paradoxical formula

$$(a \neq b) = (a = b').$$

This is easily understood, for, whatever the proposition b, either it is true and its negative is false, or it is false and its negative is true. Now, whatever the proposition a may be, it is true or false; hence it is necessarily equal either to b or to b'. Thus to deny an equality (between propositions) is to affirm the *opposite* equality.

[60]This argument is general and from it we can deduce the formula

$$\prod_x (a + x) = a,$$

whence may be derived the correlative formula

$$\sum_x ax = a.$$

Thence it results that, in the calculus of propositions, we do not need to take inequalities into consideration—a fact which greatly simplifies both theory and practice. Moreover, just as we can combine alternative equalities, we can also combine simultaneous inequalities, since they are reducible to equalities.

For, from the formulas previously established (§0.57)

$$(ab = 0) = (a = 0) + (b = 0),$$
$$(a + b = 1) = (a = 1) + (b = 1),$$

we deduce by contraposition

$$(a \neq 0)(b \neq 0) = (ab \neq 0),$$
$$(a \neq 1)(b \neq 1) = (a + b \neq 1).$$

These two formulas, moreover, according to what we have just said, are equivalent to the known formulas

$$(a = 1)(b = 1) = (ab = 1),$$
$$(a = 0)(b = 0) = (a + b = 0).$$

Therefore, in the calculus of propositions, we can solve all simultaneous systems of equalities or inequalities and all alternative systems of equalities or inequalities, which is not possible in the calculus of classes. To this end, it is necessary only to apply the following rule:

First reduce the inclusions to equalities and the non-inclusions to inequalities; then reduce the equalities so that their second members will be 1, and the inequalities so that their second members will be 0, and transform the latter into equalities having 1 for a second member; finally, suppress the second members 1 and the signs of equality, *i.e.*, form the product of the first members of the simultaneous equalities and the sum of the first members of the alternative equalities, retaining the parentheses.

0.60 Conclusion

The foregoing exposition is far from being exhaustive; it does not pretend to be a complete treatise on the algebra of logic, but only undertakes to make known the elementary principles and theories of that science. The algebra of logic is an algorithm with laws peculiar to itself. In some phases it is very analogous to ordinary algebra, and in others it is very widely different. For instance, it does not recognize the distinction of *degrees*; the laws of tautology and absorption introduce into it great simplifications by excluding from it numerical coefficients. It is a formal calculus which can give rise to all sorts of theories and problems, and is susceptible of an almost infinite development.

But at the same time it is a restricted system, and it is important to bear in mind that it is far from embracing all of logic. Properly speaking, it is only the algebra of classical logic. Like this logic, it remains confined to the domain circumscribed by Aristotle, namely, the domain of the relations of inclusion between concepts and the relations of implication between propositions. It is true that classical logic (even when shorn of its errors and superfluities) was much more narrow than the algebra of logic. It is almost entirely contained within the bounds of the theory of the syllogism whose limits to-day appear very restricted and artificial. Nevertheless, the algebra of logic simply treats, with much more breadth and universality, problems of the same order; it is at bottom nothing else than the theory of classes or aggregates considered in their relations of inclusion or identity. Now logic ought to study many other kinds of concepts than generic concepts (concepts of classes) and many other relations than the relation of inclusion (of subsumption) between such concepts. It ought, in short, to develop into a logic of relations, which LEIBNIZ foresaw, which PEIRCE and SCHRÖDER founded, and which PEANO and RUSSELL seem to have established on definite foundations.

While classical logic and the algebra of logic are of hardly any use to mathematics, mathematics, on the other hand, finds in the logic of relations its concepts and fundamental principles; the true logic of mathematics is the logic of relations. The algebra of logic itself arises out of pure logic considered as a particular mathematical theory, for it rests on principles which have been implicitly postulated and which are not susceptible of algebraic or symbolic expression because they are the foundation of all symbolism and of all the logical calculus.[61] Accordingly, we can say that the algebra of logic is a *mathematical* logic by its form and by its method, but it must not be mistaken for the logic *of mathematics*.

[61]The principle of deduction and the principle of substitution. See the author's *Manuel de Logistique*, Chapter 1, §§ 2 and 3 [not published], and *Les Principes des Mathématiques*, Chapter 1, A.

Index